从零开始学
Flutter开发

谭东●著

電子工業出版社
Publishing House of Electronics Industry
北京•BEIJING

内 容 简 介

本书针对目前高速发展的Flutter跨平台移动开发技术方案，从零开始深入讲解其中涉及的技术点，内容全面详细。本书共分18章，第1章至第4章主要介绍Flutter入门基础知识，第5章至第7章主要介绍Flutter核心组件和布局相关组件，第8章至第15章主要介绍Flutter进阶知识，第16章至第18章主要介绍Flutter扩展及实战相关内容。

本书适合具有一定编程经验的学生、开发者阅读，也适合乐于尝试新技术、渴望不断提升自我的读者参考、学习。

图书在版编目（CIP）数据

从零开始学Flutter开发 / 谭东著. —北京：电子工业出版社，2020.9
ISBN 978-7-121-38713-5

Ⅰ.①从… Ⅱ.①谭… Ⅲ.①移动终端－应用程序－程序设计 Ⅳ.①TN929.53

中国版本图书馆CIP数据核字(2020)第039500号

责任编辑：孙奇俏
印　　刷：三河市良远印务有限公司
装　　订：三河市良远印务有限公司
出版发行：电子工业出版社
　　　　　北京市海淀区万寿路173信箱　　邮编：100036
开　　本：787×980　1/16　印张：27　字数：602.2千字
版　　次：2020年9月第1版
印　　次：2020年9月第1次印刷
定　　价：108.00元

凡所购买电子工业出版社图书有缺损问题，请向购买书店调换。若书店售缺，请与本社发行部联系，联系及邮购电话：（010）88254888，88258888。

质量投诉请发邮件至zlts@phei.com.cn，盗版侵权举报请发邮件至dbqq@phei.com.cn。

本书咨询联系方式：010-51260888-819，faq@phei.com.cn。

推荐语

Flutter 自发布了 1.0 正式版之后，便迅速成长，相继在 Web 端和 PC 端开发领域也取得了较大突破。目前，全球已有越来越多的公司采用 Flutter 进行移动端应用开发，各位工程师也应跟进脚步，投入学习。本书全面介绍了 Flutter 开发涉及的知识点和工具，可以作为快速入门的教材，也可以作为实际开发过程中的参考手册。

字节跳动工程师，桑明明

Flutter 以其高性能、跨平台、高效率的特性，让传统意义上的前端开发人员得以快速入门移动端开发，事实上它正在成为新一代的移动端开发标准。本书是你学习 Flutter 路上的好伙伴，跟随作者的节奏，你将完成从 Flutter 新手到熟手的跃迁。

G7 架构师、蚂蚁金服前工程师、《重新定义 Spring Cloud 实战》作者之一，叶志远

“跨平台开发”是近年来的热门话题，虽解决方案众多，但 Flutter 尤为突出。网络上关于 Flutter 的资料比较碎片化，要想系统学习还是要参考图书。谭东为了写好这本书投入了不少心血，书中的每个知识点都是他不断学习与实践的经验总结。相信这本书能帮助初学者高效学习，少走弯路。

芒果 TV 前工程师，周展

前言

Flutter 是 Google 推出的新兴跨平台移动开发技术方案，由于其开发效率高，性能优秀，并且有 Google 的大力支持，因此发展迅速，收获了许多关注。

Flutter 不仅前期表现亮眼，其未来规划也值得我们期待。它的目标是实现移动端、Web 端、PC 端、服务器后端等平台的应用开发，成为真正的高性能、跨平台技术方案。

目前在 Google 内部，Flutter 已被广泛应用于多个产品，比如 Google Ads 产品的 iOS 版本和 Android 版本都使用 Flutter 开发。全世界也有多家大型公司开始使用 Flutter 来开发应用，包括 Abbey Road Studios、阿里巴巴、Capital One、Groupon、Hamilton、京东、腾讯等。

表现突出的 Flutter 同样赢得了许多开发者的青睐，很多开发者转型学习 Flutter 开发。在众多青睐者的努力下，Flutter 社区越来越完善。有许多热情的学习者提供了关于 Flutter 的文档、资源、第三方插件库，Flutter 官方也提供了 Dart Pub 插件库平台来帮助开发者提高开发效率。在 GitHub 最受欢迎开源软件排行榜中，Flutter 排名前 20。所以我们有理由相信，Flutter 会变得越来越好，它势必成为未来的主流跨平台开发技术方案。

基于此，我想要将自己的实际开发经验和对 Flutter 开发的心得体会，总结成书，帮助开发者和学习者从零开始，全面、细致地学习 Flutter 相关技术。这本书从大纲策划，到实际写作，再到后期内容完善，我都十分认真地对待，投入了非常多的心血。本书的内容几乎覆盖了 Flutter 开发涉及的全部知识点，体现了从零开始学习 Flutter 开发到进行实战的过程。

衷心希望大家能够认真学习 Flutter，因为对于一门新兴的、有前景的技术而言，如果你能成为第一批学习者，那么你将极有可能成为该技术的引领者，实现收获最大化。

很希望通过这本书和大家一起成长、进步，让我们一同期待 Flutter 更好的未来！

本书内容

本书内容全面，基本涵盖了 Flutter 开发涉及的所有知识点。全书共分 18 章，每章的内容简介如下。

第 1 章　认识 Flutter

Flutter 是 Google 公司推出的开源跨平台移动开发技术方案，本章将简单介绍跨平台开发技术，然后围绕 Flutter 的架构、特点、主流平台和未来展望带领读者全面认识 Flutter。

第 2 章　Dart 语言基础

学习一门编程语言，一定要了解它的特性，本章将首先介绍 Dart 语言的基础知识和特性，然后介绍 Dart 的数据类型与操作符、流程控制语句、类与方法、异步操作与导入类，为 Flutter 开发打下坚实的理论基础。

第 3 章　Flutter 开发入门

从本章开始将正式进入 Flutter 开发环节，本章内容涉及开发环境搭建、项目结构分析、配置文件详解、组件分类，以及创建 Flutter 应用。

第 4 章　Flutter 开发规范

本章将着重讲解 Flutter 开发规范，主要内容包括 Flutter 的项目结构规范、命名规范、代码格式规范、注释规范，以及代码使用规范。

第 5 章　Flutter 常用组件（上）

本章将介绍 Flutter 中的常用组件。在 Flutter 中，几乎所有的对象都可以看成组件，组件不单单是 UI 控件，也具备一些逻辑操作功能。本章将主要介绍文本类组件、图片类组件、导航类组件。

第 6 章　Flutter 常用组件（下）

在上一章内容的基础上，本章将继续介绍 Flutter 中的表单类组件、列表滚动组件、Dialog 组件、表格组件。

第 7 章 Flutter 常用布局组件

在 Flutter 中，布局也可以看作一个组件。本章将对 Flutter 常用布局组件中的典型布局组件进行讲解，并结合案例深入实践，内容涉及容器类布局、层叠类布局、线性布局、弹性布局和流式布局等组件。

第 8 章 Flutter 路由与生命周期

在 Flutter 中，路由负责页面跳转和数据传递，是非常重要的概念。本章将主要讲解 Flutter 中路由的概念、使用方法，路由跳转的实现，参数传递，按键监听，以及 Flutter 的生命周期。

第 9 章 Flutter HTTP 网络请求

Flutter 中 HTTP 网络请求的实现方法主要有三种：通过 io.dart 里的 HttpClient 实现、通过 Dart 原生 HTTP 请求库实现、通过第三方库实现。本章将详细讲解这三种方法的特点和区别，并扩展介绍 JSON 编解码和 WebSocket 的使用方法。

第 10 章 Flutter 文件操作与数据库操作

实际开发中离不开文件操作，Flutter 也提供了相关的文件操作 API，同时支持数据库操作。本章将结合实例介绍 Flutter 的文件操作、数据库操作，同时介绍 Flutter 的手势操作。

第 11 章 Flutter 自定义组件与方法封装

在开发过程中，有些需求无法通过现有的 Flutter 组件实现，这时就要自定义组件。本章将介绍 Flutter 中自定义组件的几种方式，同时也会讲解 Flutter 方法的封装。

第 12 章 Flutter 动画的实现

如果想让应用或产品的用户体验变得更好，动画效果是一个很重要的因素。本章将介绍 Flutter 中动画的基本使用方法和特点，涉及基础动画、Hero 动画、交错动画。

第 13 章 Flutter 主题与应用国际化

Flutter 中也有“主题”这一概念。国际化可以让应用支持多种语言。本章将主要介绍 Flutter 中主题的多种实现，以及应用国际化的实现。

第 14 章　Flutter 数据共享与传递

本章将配合实例详细介绍 Flutter 中数据共享与传递的方法：通过 InheritedWidget 组件、通过 ScopedModel 库、通过 Redux 库、通过 EventBus 库。同时也会介绍一些与数据交互相关的插件库。

第 15 章　Flutter 与原生 API 交互及插件库开发

在使用 Flutter 进行开发的过程中，有时需要编写插件来实现相应的交互功能。本章将介绍 Flutter 与原生 API 交互的方法，Flutter 插件库开发方法，以及常见插件库的用法，并配合实例详细说明。

第 16 章　Flutter 调试与应用打包发布

在使用不同的开发语言进行开发时，一般都会用到调试功能。Flutter 也支持调试和单元测试。本章将介绍在 Flutter 中进行调试和单元测试的方法，同时介绍 Flutter 应用打包与发布的流程。

第 17 章　Flutter 拓展：Dart Web

Dart 除了可以用于 Flutter 移动应用开发，还可以用于 Web 开发。本章将拓展介绍 Dart Web 相关开发知识，包括 Dart Web 开发环境搭建、Dart Web 项目的创建与运行等。

第 18 章　Flutter 实战

本章将基于前面章节的内容进行实战：实现一个简易备忘录应用，以及编写一个 TV 应用。通过这两个示例，读者可以巩固之前学过的知识，高效查缺补漏。

阅读准备

要想运行本书中的示例代码，需要配置和安装如下系统环境和软件。

- 操作系统：Windows、macOS、Linux 均可。
- Dart 环境：建议安装最新稳定版本的 Dart SDK。
- Flutter 环境：建议安装最新稳定版本的 Flutter SDK。
- 开发环境：Android Studio 或 Visual Studio Code 均可。

联系作者

Flutter 开发领域有许多深奥的知识，尽管我在写作过程中力求严谨，但书中仍可能存在一些疏漏，敬请各位读者指正，我会及时调整、修改。

可通过电子邮箱 852041173@qq.com 与我交流。我会将反馈信息整理在博客中，大家可通过 https://fantasy.blog.csdn.net 访问、查看。另外，也欢迎各位读者关注我的微信公众号 tandongjay，我会定期分享一些技术文章。

致谢

感谢我的父母、妻子在我写作过程中给予的支持和鼓励。

感谢我的老师让我对计算机产生了浓厚的兴趣，培养了我的相关技术思维。

感谢我的公司为我提供了实际的项目研发和技术攻关机会，我从中总结了很多经验，攻克了很多技术难题。

感谢电子工业出版社的孙奇俏老师在本书出版过程中给予我的大力支持与帮助，她非常专业、耐心。

谭东

2020 年 6 月

读者服务

微信扫码回复：38713

- 获取本书配套代码资源
- 加入本书读者交流群，与本书作者互动
- 获取博文视点学院 20 元付费内容抵扣券
- 获取更多技术专家分享的视频与学习资源

目录

第1章 认识 Flutter

Flutter 是 Google 公司于 2015 年 5 月 3 日推出的开源跨平台移动开发技术方案，可用于快速开发 Android 和 iOS 应用，也可用于开发 Web 应用，同时还是在 Google Fuchsia 下开发应用的主要工具。Flutter 的引擎使用 C++语言开发，基础库是通过 Dart 语言编写的。Dart 提供了用 Flutter 开发应用所需的基本的类和函数。

Flutter 的第一个版本运行在 Android 操作系统上，被称作“Sky”，它于 2015 年在 Flutter 开发者会议上被公布。Flutter 1.0 正式版本于 2018 年 12 月 5 日发布，功能基本完善。截止到本书写作之时，Flutter 最新版本为 1.17 正式稳定版。

话不多说，下面就带大家来认识 Flutter。

1.1 跨平台开发技术

移动开发从原始的原生应用开发，到基于 Kotiln、Swift 语言开发，再到基于各种跨平台技术开发，经历了多次演变。不过上述方案或多或少都有一些局限性，于是 Google 公司推出了新的跨平台移动开发技术方案 Flutter。

在移动平台上，原生应用的体验最好，流畅度最高，性能也最好。目前基于跨平台技术开发的应用，其流畅度和体验远远达不到原生应用的效果，多少都会卡顿、丢帧。而基于 Flutter 开发的应用在体验和流畅度上基本能媲美原生应用，不会卡顿、丢帧，官方显示其渲染性能可以达到 120 FPS。

由于 SDK 里所有的布局、控件都被组件化，因此使用 Flutter 开发应用时效率非常高。Flutter 不仅仅局限于跨平台移动应用开发，还支持 Web 开发、后端开发、PC 桌面应用开发（内测中）、嵌入式开发（内测中），这也是 Flutter 越来越受到关注，被越来越多的大公司和开发者使用的原因之一。

Flutter 支持多种开发工具的插件化使用，能契合不同开发者的习惯，同时做到一套代码逻辑跨平台运行。一些与原生交互的代码可以通过插件形式使用，依然能兼容多个平台。

目前我们在开发应用时，要想兼容 iOS 和 Android 两个平台，可以有两种技术选择：走原生开发路线，在不同平台上分别实现界面和逻辑；用同一套代码兼容多个平台，但这往往意味着运行速度和用户体验的损失。Flutter 的出现为我们提供了一套两全其美的解决方案：既能直接调用原生代码来加速图形渲染和 UI 绘制，又能同时运行在两大主流移动平台上，用户体验和运行速度与原生开发基本一致，效率非常高，学习难度也较低。接下来我们来对比几种跨平台开发技术，如表 1-1 所示。

表 1-1　跨平台开发技术对比

技　术	性　能	开发效率	渲染方式	学习成本	可扩展性
Flutter	好，接近原生体验	高	Skia 高性能自绘引擎	低，组件化	高，采用插件化的库进行扩展
React Native/Weex/小程序	一般，有延迟	一般	JavaScript 驱动原生渲染	高，复杂	一般
原生开发方式	好	一般	原生渲染	高，需要学习 Android 和 iOS 原生 API	高
HTML5	差	一般	HTML/JavaScript 渲染	高，需要学习 HTML、CSS 和 JavaScript	一般

从表 1-1 中可以看出，Flutter 的优势明显，具体来说如下。

- 跨平台，一套代码可以运行在 Android 和 iOS 平台上，未来还可以运行在 Fuchsia OS 平台上。
- 接近原生应用的用户体验和性能。
- 开发快速，不用单独写布局文件，直接组合相关组件和配置属性。
- 毫秒级的热重载，修改后可以立即看到效果，开发和调试非常高效。
- 自绘引擎，不依赖于 WebView 渲染，性能好，体验好。

早在 Flutter 1.0 正式版本尚未推出之前，就已经有成百上千的基于 Flutter 开发的应用在 Apple Store 和 Google Play 上架，相信 Flutter 将会被越来越多的开发者和公司所采用。

1.2　一起认识 Flutter

目前，全球已有多个公司使用 Flutter 开发应用，其更新频率之高，让更多的开发者和公司看到了希望。Flutter 势必成为未来的跨平台开发主流技术。下面我们来简单回顾一下近年来 Flutter 的发展情况。

- 2018 年 2 月，第一个 Beta 版本发布。
- 2018 年 5 月，Google I/O 大会上发布了 Beta 3 版本。
- 2018 年 6 月，GMTC 上发布了首个预览版本。
- 2018 年 9 月，Google 开发者大会上发布了第二个预览版本。
- 2018 年 12 月，正式稳定的 1.0 版本发布。
- 2019 年 2 月，稳定的 SDK 1.2 版本发布。
- 2019 年 5 月，稳定的 SDK 1.5 版本发布。
- 2019 年 9 月，稳定的 SDK 1.9 版本发布。
- 2019 年 12 月，稳定的 SDK 1.12 版本发布。
- 2020 年 5 月，稳定的 SDK 1.17 版本发布。

2019 年，Flutter 经历了几次版本迭代，其中 Flutter SDK 1.12 稳定版围绕以下内容进行了优化与改进。

- 提升了核心框架的稳定性、性能和质量。
- 改进了现有组件的视觉效果和功能。
- 为开发者提供了全新的基于 Web 的调试工具。

Flutter SDK 1.17 稳定版围绕以下内容进行了优化与改进。

- 优化了移动端性能和文件体积。
- 新增了 Material 组件：NavigationRail、DatePicker 等。

- 增加了对文字缩放、排版的支持，并新增了一些动画效果。
- 支持使用 Google Fonts。
- 新增了无障碍功能和应用国际化的扩展升级。
- 提供了新的 Dart DevTools，辅助开发调试。

Flutter 是基于 Dart 语言编写的，语言风格和 React 很像。Flutter 应用及其功能几乎都是采用组件来构建的，组件采用现代响应式框架构建。可以说，一切对象都是组件。

用 Flutter 开发应用的效率相对于原生开发提升了很多，其基本功能也已经很完善，可以说 Flutter 实现了技术突破，Dart 未来也有望成为热门主流语言。国内最早把 Flutter 应用于商业项目的是阿里巴巴的闲鱼团队。

简单了解了 Flutter，接下来我们来看一下编程语言 Dart。

Dart 是 Google 公司开发的计算机编程语言。2011 年 10 月 10 日，在丹麦奥尔胡斯举行的 GOTO 大会上，Dart 被推出，后来被 ECMA 认定为标准。Dart 非常强大，目前可以用于 Web 应用、移动应用、物联网应用开发等，是真正的高性能、跨平台开发语言。Dart 是面向对象的结构化编程语言。

我们知道，任何技术和语言的发展与它的社区支持密不可分。Flutter 已经被很多大公司采用，具体内容可以在 Flutter 官方网站查看。这里截取了采用 Flutter 的部分公司列表，如图 1-1 所示。

目前 Flutter 社区非常活跃，Flutter 也位列 GitHub 最受欢迎的开源软件前 20 名。开发者可以关注 GitHub 上 Flutter 项目的更新动态，官方更新频率很高，这一点非常好，对开发者非常有帮助。 遇到问题时除了可以使用搜索引擎进行搜索，还可以在官方 GitHub 上进行提问。

另外，我们可以在官方仓库查找第三方插件库并使用。Flutter 官方开源插件仓库为 Dart Pub，其界面如图 1-2 所示。

Dart Pub 里面有非常多的插件供我们使用，还配备有相关文档，当然我们也可以提交自己的开源插件到其中。

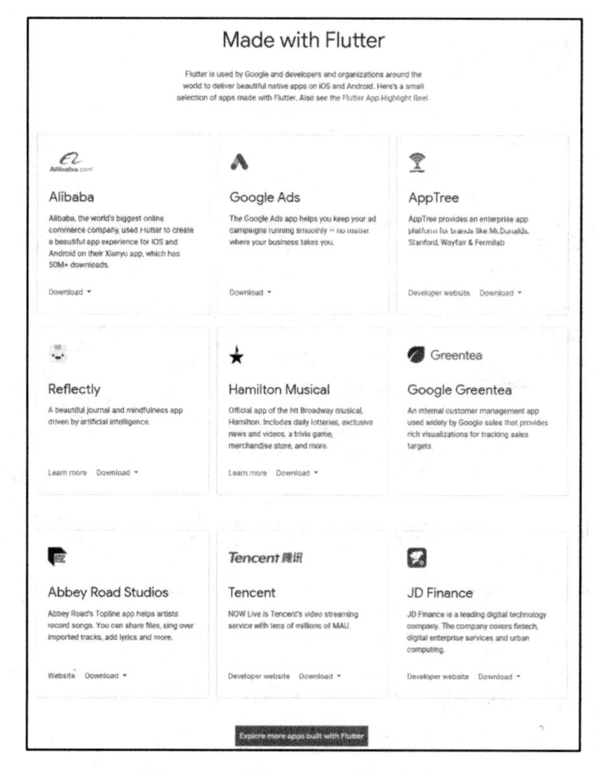

图 1-1　采用 Flutter 的部分公司

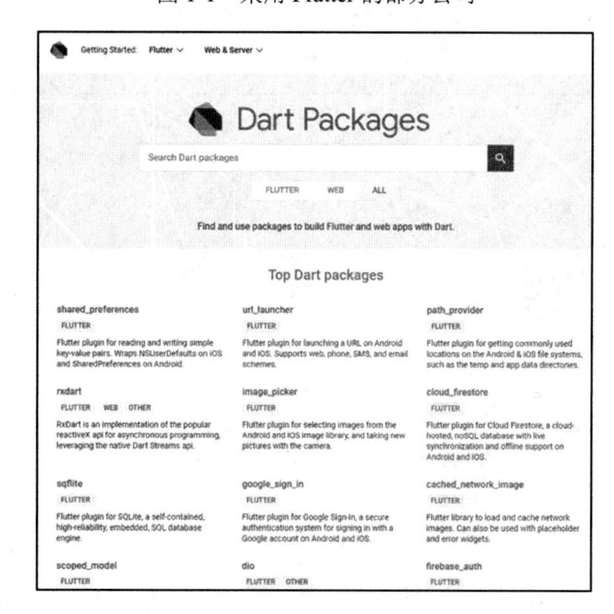

图 1-2　Dart Pub 界面

1.3 Flutter 架构与特点

Flutter 提供了两种风格的 UI 组件来分别适应 Android 应用和 iOS 应用的特点，分别是 Material 风格组件和 Cupertino 风格组件。Flutter 的开发特点是，组件是由许多更小的组件组合而成的，类的层次结构也是扁平的。官方给出的 Flutter 架构如图 1-3 所示。

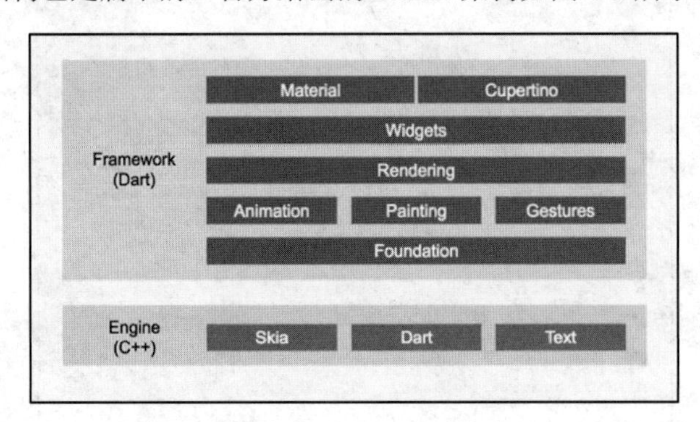

图 1-3 官方给出的 Flutter 架构

从图 1-3 中可以看出，核心引擎（Engine）是使用 C++语言编写的，Framework 层采用了由 Dart 语言编写的 SDK，并且提供了 Material 和 Cupertino 两套不同风格的 UI 组件。

Flutter 其实很像一门胶水语言，我们可以用 Dart 语言编写一套应用，然后它会被编译、调用或渲染成不同平台的应用。我们来看一下 Flutter 的组件分类，如图 1-4 所示。

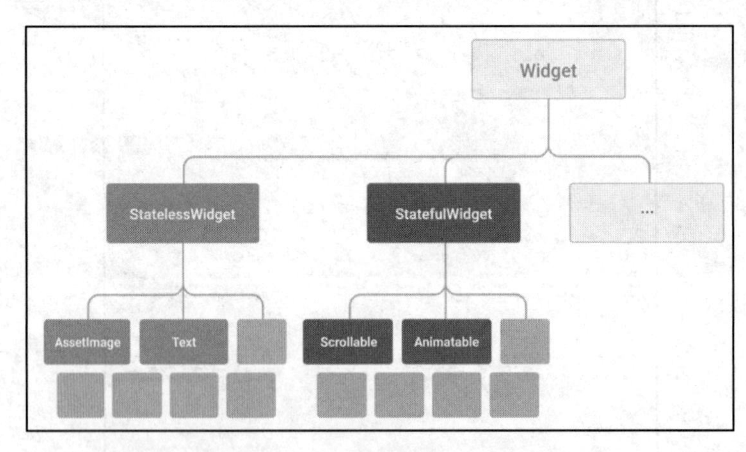

图 1-4 Flutter 的组件分类

Flutter 中的一切都可以看成组件（Widget），主要分为有状态组件（StatefulWidget）和无状态组件（StatelessWidget）两类。StatelessWidget 主要用于不需要维护状态的场景，StatefulWidget 主要用于需要维护好状态的场景。

前面我们简单介绍了 Flutter 的架构与组件分类，接下来我们来详细了解 Flutter 的特点，具体如下。

- 运行速度快：媲美原生应用运行速度，用户体验很好。
- 高效：引入 Stateful Hot Reload（保持应用状态的热重载），可以让移动开发者和设计师们实时快速预览应用程序。通过 Stateful Hot Reload，无须重新启动应用就可以在程序运行时直接看到代码修改之后的效果。
- 开源：Flutter 是一个基于 BSD-style 许可的开源项目，全球数百位开发者为之贡献代码，Flutter 的插件生态系统平台资源非常丰富，有数千款插件已经发布，避免了重复造轮子。
- 开放：可以使用原生资源开发应用，比如我们依然可以在 Android 上使用 Kotlin 或 Java 进行开发，在 iOS 上使用 Swift 或 Objective-C 来编写逻辑。
- 提供了两套不同风格的 UI 组件。
- 支持多种开发工具：支持 Visual Studio Code、Android Studio、IntelliJ 及其他开发工具，只需要安装相关插件即可使用。

1.4　Flutter 主流平台

Flutter 官方团队始终在不断探索，将 Flutter 拓展到手机端以外的更多平台，如 Web 端、PC 端、嵌入式平台等，实现真正的跨平台—— 一套代码规范多平台运行。事实上，Flutter 的设计理念就是使其成为一个灵活且便携的 UI 工具包，以适应各种需要绘制屏幕内容的平台。

未来的 Flutter 将全面跨主流平台：Android、iOS、Windows（研发中）、macOS（研发中）、Linux（研发中）、Fuchsia OS（研发中）、Web（已经发布测试版 SDK）、物联网系统（研发中）、后端等。下面我们大致介绍一下这些主流平台 SDK 的研发情况。

1. Fuchsia OS

Google 的 Fuchsia OS 可谓未来的热门操作系统，它的特点吸引了一大批开发者和学习者。Fuchsia OS 默认支持基于 Flutter 和 Dart 开发的应用。

Fuchsia OS 是一套可以运行在手机端、PC 端等不同平台上的跨平台系统，放弃了 Linux 内核，基于 Zircon 微核，采用 Flutter 引擎和 Dart 语言编写。Fuchsia OS 正式版预计会在 2021 年推出，其或许会替代 Android 系统。据传，Google 聘请了有着多年 macOS 开发经验的资深工程师来负责 Fuchsia OS 开发，目标是将其推向市场。

2．Flutter Web

Flutter Web 是一个基于 Web 技术原理实现的 Flutter 运行时环境。通过对 Web 端的支持，Flutter 应用程序无须改动就能运行在标准 Web 平台上。Flutter Web 可以大大简化我们开发 Web 页面的操作，无须编写繁杂的 CSS、JavaScript、HTML，一套 Flutter 代码就能轻松搞定一个 Web 页面系统。

虽然 Flutter Web 正式版 SDK 还没有发布，但是已经有了预览版本可以体验，如图 1-5 所示。

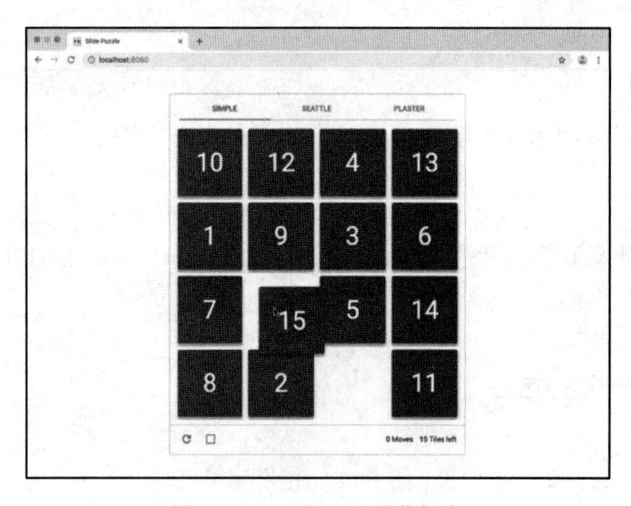

图 1-5　Flutter Web 预览版本

目前，Flutter Web 应用在桌面浏览器端基本能达到每秒 60 帧的渲染速度，但是在移动浏览器端，特别是在低端机型上还有很大的优化空间。

3．Flutter 桌面和嵌入式

Flutter 也将支持桌面 PC 平台，目前处于研发阶段。未来可以用 Flutter 开发能够在 Windows、Linux、macOS 上运行的应用。目前 Flutter 已经推出了桌面应用和嵌入式应用的小例子，体验还不错，大家可以去 Flutter 官网和 GitHub 上学习。

目前 Flutter 桌面和嵌入式项目处于实验测试阶段，相信在不久的将来会发布可用的 SDK 版本。

4. Flutter 游戏

Flutter 团队和 2Dimensions 联合发布了一款游戏 Flutter Developer Quest。这款游戏是完全通过 Flutter 开发构建的，不仅在游戏性上毫不缩水，其代码也完全开源。可以说，这是一项 Flutter 在游戏开发领域的新尝试和应用拓展，该游戏的源代码可以在 GitHub 上获得。

1.5 Flutter 未来展望

Flutter 和 Dart 目前依然在不断迭代、进步。相信在不久的将来我们可以看到一些关于 Flutter 和 Dart 的新内容。

根据 GitHub 上公布的 Flutter 规划路线，我们要着重关注核心和基础、易用性、生态系统、移动端之外的支持、动态更新、工具链。当然，我们也可以提出其他反馈意见。

1. 核心和基础

Flutter 关注的核心和基础内容有以下几个方面。

- 修复 Bug：优先级主要基于 Issue 下的互动数量。
- 性能调优：包括减少内存和引擎占用空间，提高帧率等。
- 改进 Flutter 工程测试流程：确保为开发者们提供稳定的版本，不会出现版本回归。
- 改进错误提醒：使错误提醒更具可操作性，提供一些常见的解决方案。
- API 文档改进：提供示例代码和图表等，让 API 文档更易用。

2. 易用性

在易用性方面，Flutter 关注的重点内容如下。

- 满足希望使用混合工程（将 Flutter 集成到现有的 Native 工程项目中）的开发者们的需求，如提供新的插件模板和 Android 内嵌 API。
- 更新 Flutter 官方文档以提供更详尽的使用教程。

- 在 Flutter 应用里管理状态。
- 投入时间持续更新和维护 Cupertino 组件。
- 使 Flutter 在未安装完整工具链和运行环境前更容易体验和使用。

3．生态系统

Flutter 同时也在完善自己的生态系统，具体体现如下。

- 获得更好的 C、C++库支持，包括 Dart、C、C++之间的相互调用。
- 推进官方开发、维护的 Packages，达到与核心框架代码相同的质量和完整性。
- 在 iOS 和 Android 上完成地图和 WebView 插件的开发。
- 确保应用可以使用一些 Google 服务，比如在应用内支付。
- 提供本地推送通知，支持本地数据存储。

4．移动端之外的支持

通过以下做法将 Flutter 拓展到更多终端平台，可以实现“构建一个便携的 UI 工具包，在任何需要的地方画出每一帧像素”的目标。

- 更好地支持键盘和鼠标的输入。
- 实现 Hummingbird 项目，让 Flutter 可以运行在 Web 平台。
- 让 Flutter 可以运行在桌面平台（如 Windows）上。

5．动态更新

Dart 语言为 Flutter 应用开发提供了热更新（dynamic update）特性，开发者们无须重新部署就可以把代码推送到应用中。

- Android 上的动态修复：让开发者直接将代码更新从服务器端推送到 Android 应用中。
- 动态载入：让应用里不常用的部分延迟加载。

6．工具链

要继续投入精力支持 Visual Studio Code、Android Studio 和 IntelliJ，使它们能够成为开发

Flutter 的主要 IDE。

- 增加对 Language Server Protocol 及其他开放协议的支持。
- 通过改进开发过程中的分析、调试体验，提高应用的整体质量和性能。
- 持续优化应用模板，让 Flutter 上手更快。

Flutter 现已进入 GitHub Top20 软件库，成为未来主流跨平台开发技术势在必行。它的高效、开放、一套代码多终端运行等特点都非常吸引人。

在这一章中，我们简单地带大家认识了 Flutter，希望大家可以对 Flutter 有正确的认识和了解，并且对于学习 Flutter 充满信心。接下来就让我们一起入门 Flutter 开发，掌握未来主流技术的主动权吧！

第2章 Dart语言基础

学习一门编程语言，首先要了解它的历史和特点。

本章将主要介绍 Dart 的基础知识和特性、数据类型、操作符、流程控制语句、类与方法、异步操作与导入类等。Dart 语法很简单，有了 Dart 基础，学习 Flutter 将会更快、更容易。

2.1 认识 Dart

Flutter 是基于 Dart 语言编写的，所以 Flutter 中的语法基于 Dart 语法，学习 Flutter 就要先了解 Dart。

2.1.1 什么是 Dart

Dart 是 Google 公司推出的编程语言，于 2011 年正式亮相。Dart 是一门面向对象编程语言，语法和 Java、C、JavaScript 语言很像。所以会 Java 语言的人，学习 Dart 一般会快一些。Dart 里所有的类都可以看成对象，它是单继承、动态类语言。Dart 可用于多平台开发（见图 2-1），目前还在不断扩展开发平台，可以说 Dart 在各个平台领域“无所不能”。

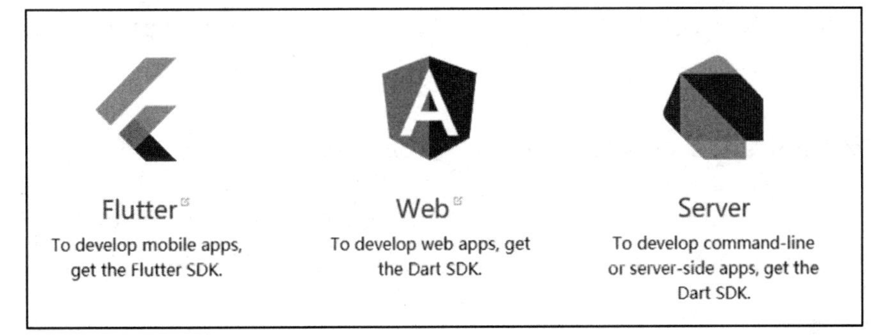

图 2-1　Dart 所支持的开发平台

Dart 可以说集合了 Java 和 JavaScript 的一些特性和优点。在静态语法方面，如类型定义、函数声明等，Dart 和 Java 非常相似。在动态特性方面，如函数式特性、支持异步等，Dart 又和 JavaScript 很像。除融合了 Java 和 JavaScript 的优点外，Dart 也具有一些自身语法特点，如命名参数可选、支持级联运算符等。

2.1.2　Dart 的特性

Dart 有很多优点，我们可以根据 Dart 的一些特性来选择是否需要使用它，具体如下。

- 语法简单明了，开发速度快、效率高，学习成本低。
- 功能强大，可以开发 Web 端、移动端、PC 端、服务器端、物联网等平台应用。
- 编译、执行速度快，拥有自己的 Dart VM。
- 可移植，类似于中间件语言，可以编译成不同平台的原生代码，方便扩展成跨平台应用语言。
- 融合了 Java、C、JavaScript 语言的特点，结合了 React 响应式编程的思想规范。

在实际开发时，主要是按照语法规范来编写代码逻辑的，所以我们需要同时了解 Dart 的语法规范及语言本身的特性，这样更利于我们进一步了解 Dart 的使用规则。

- 面向对象语言，一切数据类型、API 都是对象，都继承自 Object 类。
- 强类型语言，同时也是动态类型语言。
- 没有设置定义访问域的关键字，如果某个变量、方法、类的名称以下画线“_”开头，则说明这个变量、方法、类是私有的，外部不可以调用。

- 有入口函数 main(){...}，类似于 Java 中的 public void main(String[] args){...}。
- 明显缩减了代码量，提高了可读性。
- 支持 Future 和 Streams 使用方式，可以以类似 RxJava 式的方式使用。

接下来我们看一下 Dart 中的关键字（33 个保留关键字，17 个内置标识符，6 个 Dart 2 新增异步功能关键字，25 个 Dart 特有关键字）。

33 个 Dart 保留关键字如表 2-1 所示。

表 2-1　33 个 Dart 保留关键字

保留关键字						
assert	break	const	continue	case	catch	class
default	else	enum	extends	final	finally	false
for	if	in	is	new	null	rethrow
return	superdo	switch	throw	try	typedef	this
true	var	void	while	with		

17 个 Dart 内置标识符如表 2-2 所示。

表 2-2　17 个 Dart 内置标识符

内置标识符						
abstract	as	covariant	deferred	dynamic	export	external
factory	get	implements	import	library	operator	part
set	static	typedef				

6 个 Dart 2 新增异步功能关键字如表 2-3 所示。

表 2-3　6 个 Dart 2 新增异步功能关键字

异步功能关键字						
async	async*	await	sync*	yield	yield*	

25 个 Dart 特有关键字（和 Java 相比）如表 2-4 所示。

表 2-4　25 个 Dart 特有关键字

特有关键字						
as	assert	async	async*	await	const	covariant
deferred	dynamic	export	external	factory	get	in
is	library	operator	part	rethrow	set	sync*
typedef	var	yield	yield*			

对于以上关键字，我们不可以使用与之相同的名称为变量或常量命名，并且需要注意这些关键字的含义和作用。

2.2　Dart 的数据类型与操作符

接下来我们学习 Dart 语言基础。我们知道，学习任何一门编程语言都要知道它的数据类型和操作符有哪些，有什么特点。本节我们将对 Dart 的数据类型和操作符进行讲解。

2.2.1　Dart 中的数据类型

本节将主要介绍 Dart 的数据类型。在介绍数据类型之前，我们先来介绍 Dart 中变量的声明方法。

Dart 中变量的声明

Dart 中变量的声明方法和其他语言中的大同小异，不过也有一些自己的特点，常见的声明方法有以下三种。

1．使用 var 关键字声明

在 Dart 中，使用 var 关键字声明变量与 JavaScript 中的相似，var 可以接收任何类型的变量。但最大的不同是，Dart 中的 var 变量一旦被赋值，其类型便会确定，不能再改变，示例如下。

```
var name;
name = "hello world";
// 如果改变类型就会报错，像下面这样声明会报错
name = 900;
```

Dart 中的任何变量都是有确定类型的，在使用 var 关键字声明一个变量后，Dart 在编译时

会根据第一次赋值的数据类型来推断变量类型，所以更改类型在 Dart 中会报错。

2．使用 dynamic 和 Object 关键字声明

Object 是 Dart 所有对象的基类，Dart 中的所有对象、类型都是 Object 的子类，所以任何类型的数据都可以赋给由 Object 声明的变量。dynamic 与 var 一样，用它声明的变量可以由任意类型的对象赋值。dynamic、Object 与 var 的区别是，用 dynamic 和 Object 关键字声明的变量的类型在后期可以改变，示例如下。

```
dynamic x;
Object y;
x = "hello dynamic";
y = 'hello Object';
// 改变类型是可以的
x = 900;
y = 900;
```

dynamic 与 Object 的区别是，对于用 dynamic 关键字声明的变量，编译器会明确需要执行的操作，无须进行类型检测。当我们调用这个变量中不存在的方法时，编译器不会报错。而用 Object 关键字声明的变量只能使用 Object 定义的属性与方法，否则编译器会报错，示例如下。

```
dynamic x;
Object y;
main() {
   x = 3;
   y = 6;
   printLengths();
}

printLengths() {
   // 不会报错
   print(x.length);
   // 报错警告
   print(y.length);
}
```

3．使用 final 和 const 关键字声明

Dart 中 final 和 const 关键字的用法和其他语言中的类似，主要用于定义常量。其实常量是特殊的变量，只不过这个变量不会被更改。由 final 关键字声明的常量和由 const 关键字声明的

常量的区别在于，const 常量是一个编译时常量，而 final 常量是一个运行时常量，会在第一次使用时被初始化。通过 final 或者 const 关键字声明的常量，其类型声明可以省略，示例如下。

```
// 可以省略 String 类型声明
final str = "hello world";
// final String str = " hello world";
const str1 = " hello world";
// const String str1 = " hello world";
```

Dart 数据类型

了解了声明方法，我们再来了解 Dart 的数据类型。Dart 支持几种基本数据类型：Number 类型、String 类型、Boolean 类型、List 类型、Map 类型、Runes 类型、Symbol 类型。

1. Number 类型

Number 类型表示数值型数据类型，包括 int 类型和 double 类型两种。num 是 int 和 double 类型的父类。int 类型的数值范围是$-2^{53} \sim 2^{53}-1$；double 表示 64 位双精度浮点数。定义 Number 数据类型的示例如下。

```
void main() {
  // 定义 int 和 double 类型
  int a = 6;
  double b = 3.18;
  print('$a ,$b');
```

执行以上代码，输出结果如下。

```
6 ,3.18
```

2. String 类型

大家应该都很熟悉 String 类型，即字符串类型。Dart 中的 String 类型具有如下特点。

- 字符串是 UTF-16 编码的字符序列，可以使用单引号或双引号来定义。
- 可以在字符串中使用表达式，比如${expression}。
- 可以使用“+”操作符把多个字符串连接为一个字符串，也可以将多个带引号的字符串挨着写以达到同样的效果，更推荐第二种方式。

- 使用 3 个单引号或双引号可以定义多行字符串。
- 使用 r 前缀可以定义一个原始字符串。

以下是一个对 String 类型数值进行操作的示例，大家可以在这个示例中学会很多操作方法。

```
void main() {
  // 用单引号和双引号定义
  String singleString = 'A singleString';
  String doubleString = "A doubleString";
  print('$singleString ,$doubleString');

  // 使用$字符引用变量，使用{}引入表达式
  String userS = 'It\'s $singleString';
  String userExpression = 'It\'s expression,${singleString.toUpperCase()}';
  print('$userS');
  print('$userExpression');

  // 使用引号拼接字符串，或使用“+”连接字符串（不推荐）
  String stringLines =
      'String ' 'concatenation' " works even over line breaks.";
  String addString = 'A and ' + 'B';
  print('$stringLines');
  print('$addString');

  // 使用 3 个单引号或双引号来定义多行字符串
  String s3 = '''
 You can create
 multi-line strings like this one.
 ''';
  String s33 = """This is also a
 multi-line string.""";
  print('$s3');
  print('$s33');

  // 使用 r 前缀定义一个原始字符串
  String s = r"It is a \n raw string.";
  print('$s');
}
```

执行以上代码，输出结果如下。

```
A singleString ,A doubleString
It's A singleString
It's expression,A SINGLESTRING
String concatenation works even over line breaks.
A and B
You can create
multi-line strings like this one.
This is also a
multi-line string.
It is a \n raw string.
```

3. Boolean 类型

Boolean 类型即布尔类型，Dart 中用 bool 定义 true 或 false 数据类型，用法很简单。需要注意的是，有些写法在 Dart 里不被支持，将其他编程语言里 Boolean 类型的用法直接用在 Dart 里是行不通的。以下是一个在 JavaScript 里被支持，但在 Dart 里不被支持的用法示例。

```
var name = 'Tom';
if (name) {  print('He is Tom!');
}

if (1) {
  print('A line Data.');
} else {
  print('A good Data.');
}
```

4. List 类型

Dart 里使用 List 类型来表示数据集合结构。下面我们来看一下 Dart 中 List 类型的用法，示例如下。

```
void main() {
  // 定义 list
  var list = [1, 2, 3];
  List listData = [5, 6, 7];
  print(list.length);
  print(list[0]);
  // 集合数据赋值
  listData[1] = 8;
```

```
  print(listData[1]);
  // 如果在集合前加了 const 关键字，集合数据不可以操作
  var constantList = const [1, 2, 3];
  List datas = List();
  datas.add('data1');
  datas.addAll(['data2', 'data3', 'data4', 'data5', 'data6']);
  // 获取第一个元素
  print(datas.first);
  // 获取最后一个元素
  print(datas.last);
  // 获取元素的位置
  print(datas.indexOf('data1'));
  // 删除指定位置元素
  print(datas.removeAt(2));
  // 删除元素
  datas.remove('data1');
  // 删除最后一个元素
  datas.removeLast();
  // 删除指定范围中的元素，含头不含尾
  datas.removeRange(0, 2);
  // 删除指定条件的元素
  datas.removeWhere((item) => item.length > 3);
  // 删除所有的元素
  datas.clear();
  // 其他方法可以自己尝试
}
```

5. Map 类型

Dart 中的 Map 类型用于存储键/值对（key-value），键是唯一的，值允许重复。下面我们来看一下 Dart 中 Map 类型的具体用法。

```
void main() {
  var gifts = {
    // Keys  Values
    'first': 'dog',
    'second': 'cat',
    'fifth': 'orange'
  };
```

```
  var nobleGases = {
    // Keys  Values
    2: 'a',
    10: 'b',
    18: 'b',
  };
  // 定义 map 并赋值
  Map map = Map();
  map['first'] = 'a-value';
  map['second'] = 'b-value';
  map['fifth'] = 'c-value';
  Map nobleGasesMap = Map();
  nobleGasesMap[2] = 'a-value';
  nobleGasesMap[10] = 'b-value';
  nobleGasesMap[18] = 'c-value';
  // 指定键/值对类型
  var nobleGases = new Map<int, String>();
  // 获取某个键对应的值
  print(map['first']);
  // 获取 map 的大小
  print(map.length);
  // 定义一个不可变的 map
  final constantMap = const {
    2: 'a',
    10: 'b',
    18: 'c',
  };
  // 其他 API 用法和 List 类型的类似
}
```

6．Runes 类型

Runes 类型用于声明 Unicode 编码字符串（UTF-32 code points 等）。下面我们通过一个示例来看一下 Dart 中 Runes 类型的常见操作方法。

```
void main() {
  var clapping = '\u{1f44f}';
  print(clapping);
  print(clapping.codeUnits);
  print(clapping.runes.toList());
```

```
  // 用 Runes 来声明一个 Unicode 编码字符串
  Runes input = new Runes(
      '\u2665  \u{1f605}  \u{1f60e}  \u{1f47b}  \u{1f596}  \u{1f44d}');
  // 转为 String 明文字符串输出
  print(new String.fromCharCodes(input));
}
```

通过上面这段代码，我们就可以声明 Unicode 编码字符串，并转为正常的字符串打印输出。执行以上代码，结果如图 2-2 所示。

图 2-2　Runes 类型示例运行结果

7. Symbol 类型

Symbol 类型使用 Symbol 字面量来获取标识符的 symbol 对象，即在标识符前面添加一个"#"符号，示例如下。

```
#radix
#bar
```

2.2.2　Dart 中的操作符

Dart 中的操作符有很多，比如加、减、乘、除、位移等运算操作符。Dart 的运算操作符和其他语言的运算操作符用法基本一样，大家可以对比着学习。这里列出 Dart 中的操作符，如表 2-5 所示。

表 2-5　Dart 中的操作符

描　述	操 作 符
一元后缀	expr++　expr--　()　[]　.　?.
一元前缀	-expr　!expr　~expr　++expr　--expr
乘除法运算	*　/　%　~/
加减法运算	+　-

续表

描　　述	操 作 符
位操作	<<　>>
按位与	&
按位异或	^
按位或	|
比较和类型测试	>=　>　<=　<　as　is　is!
等价	==　!=
逻辑与	&&
逻辑或	||
是否为空	??
条件运算	expr1 ? expr2 : expr3
级联运算	..
赋值	=　*=　/=　~/=　%=　+=　–=　<<=　>>=　&=　^=

2.3 Dart 的流程控制语句

Dart 中的流程控制语句不多，并且比较简单，主要有以下几种。

- if 和 else
- for
- while 和 do-while
- break 和 continue
- switch 和 case
- assert 断言（判断是否相等）

Dart 也支持异常的捕获和处理等相关操作，需要用到 try-catch、throw 和 finally，可能会影响一些流程控制的跳转。下面我们通过一段整合的代码，来看一下这几个流程控制语句的基本用法。

```
void main() {
  // if 和 else
```

```
if (hasData()) {
  print("hasData");
} else if (hasString()) {
  print("hasString");
} else {
  print("noStringData");
}

// for
var message = new StringBuffer("Dart is good");
for (var i = 0; i < 6; i++) {
  message.write(',');
}

// while
while (okString()) {
  print('ok');
}
// do-while
do {
  print('okDo');
} while (!hasData());

// break 和 continue
while (true) {
  if (noData()) {
    break;
  }
  if (hasData()) {
    continue;
  }
  doSomething();
}

// switch 和 case
var command = 'OPEN';
switch (command) {
  case 'A':
    executeA();
```

```
      break;
    case 'B':
      executeB();
      break;
    case 'C':
      executeC();
      break;
    default:
      executeUnknown();
  }

  // assert 断言
  assert(string != null);
  assert(number < 80);
  assert(urlString.startsWith('https'));
}
```

Dart 也支持 Exceptions 类型的异常处理，通常使用 throw 抛出异常，示例如下。

```
throw new FormatException('Expected at least 2 section');
```

当然也可以抛出其他类型对象的异常，示例如下。

```
throw 'no data!';
```

Dart 中还可以进行异常捕获，通常使用 catch 来捕获异常，示例如下。

```
try {
  getData();
} on OutOfLlamasException {
  sendData();
} on Exception catch (e) {
  print('Unknown data Exception: $e');
} catch (e) {
  print('Some Exception really unknown: $e');
}
```

Dart 中使用 rethrow 可以将捕获的异常重新抛出，这样就可以让程序的其他部分继续捕获异常并处理，示例如下。

```
final foo = '';
```

```
void misbehave() {
  try {
    foo = "You can't change a final variable's value.";
  } catch (e) {
    print('misbehave() partially handled ${e.runtimeType}.');
rethrow;
// 使用 rethrow 重新抛出异常，允许 main()里的函数继续捕获异常
  }
}

void main() {
  try {
    misbehave();
  } catch (e) {
    print('main() finished handling ${e.runtimeType}.');
  }
}
```

Dart 中 finally 的用法和 Java 中的类似。不管是否出现异常，始终执行的方法需要写在 finally 中，示例如下。

```
try {
  getData();
} catch(e) {
  print('Error: $e');
} finally {
  // 始终执行
  sendData();
}
```

Dart 的流程控制语句非常简单，和 C++、Java 等语言中的含义和用法基本一样，大家可以对比着进行学习。

2.4　Dart 中的类与方法

类是面向对象编程语言中的概念，Dart 也是一门面向对象的编程语言，其中的类和对象与 Java 中的类和对象非常相似，也有继承和重写的特性，每个对象都是一个类的实例，都继承自 Object 对象。但 Dart 中的类又有自己的特点，本节我们就来介绍 Dart 中的类和方法。

2.4.1　Dart 中的类

Dart 中的类和 Java 等面向对象语言中的类的用法基本一样，特点也类似。不过 Dart 中的类支持多继承，并且在使用时增加了一些自己的特色，使得程序设计非常便捷。

类的创建

Dart 中使用 class 关键字来表示一个类。在类中我们可以编写各种功能函数。

创建类可以使用 new 关键字，也可以不使用。具体来说，对于 Dart 2 及以后的版本，在创建类时可以不使用 new 关键字。以下是创建类的示例。

```
class Student {
  int id;
  int age;
  String name;
}

void main() {
  // 使用 new 关键字创建类
  var student = new Student();
  // 不使用 new 关键字创建类
  var student = Student();
  student.age = 18;
  print(student.age);
}
```

如果我们想获取一个类实例的类型，在 Dart 中可以使用 Object 的 runtimeType 属性来实现，示例如下。

```
void main() {
  int a = 10;
  var b = 'Tom';
  var c = Student();

  print(a.runtimeType);// a 的类型是 int
  print(b.runtimeType);// b 的类型是 String
  print(c.runtimeType);// c 的类型是 Student
}
```

类方法的调用和类对象的赋值

Dart 中调用方法的形式是“对象.方法”，示例如下。

```
class Student {
  int id;
  int age;
  String name;

  int getRank() {
    return 20;
  }
}

void main() {
  var student = Student();
 // 通过“对象.方法”来调用类中的方法
  var rank = student.getRank();
}
```

Dart 中还有一个比较实用的方法，就是可以通过“对象?.方法”的形式来实现类方法的调用，这样可以避免当对象为 null 时程序出现异常，示例如下。

```
void main() {
  var student = Student();
  // 使用“对象?.方法”可以避免当 student 对象为 null 时程序出现异常
  var name = student?.name;
  print(name);
}
```

当然我们也可以对类对象执行赋值、取值等操作，示例如下。

```
void main() {
  var student = Student();
  // 为类对象赋值
  student.name = 'Lilei';
  // 取值
  print(student.name);
}
```

类的继承与重载

Dart 支持类的继承与重载，通过 extends 来继承父类，进而实现类的功能拓展和复用，子类可以重载父类方法，通过 super 来引用父类方法，示例如下。

```
class Product {
  void open() {
    // ...
  }
  // ...
}

// 继承与重载 Product 类
class SmartProduct extends Product {
  @override
  void open() {
    super.open();
    // 重新编写，加入新的逻辑
  }
  // ...
}
```

在以上代码中，我们使用@override 注解来表示重写父类方法，当然除了@override 注解，Dart 中还有其他注解，例如可以使用 @proxy 注解来避免警告信息，大家可以自行尝试。

除了支持单继承，Dart 还支持多继承，即可以让一个类对象继承多个父类的属性、方法，这也是 Dart 语言的一个特色。我们可以使用 with 关键字实现多继承，示例如下。

```
class Student {
  int id;
  int age;
  String name;

  int getRank() {
    return 20;
  }
}

class Product {
  void open() {
```

```
    // ...
  }
  // ...
}
// 继承 Product 类和 Student 类
class SmartProduct extends Product with Student {
  @override
  void open() {
    super.open();
    // 编写方法逻辑
  }
}

void main() {
  var product = SmartProduct();
  // 调用父类 Student 的 name 属性
  var name = product.name;
  // 调用父类 Product 的 open 方法
  product.open();
  // 调用父类 Student 的 getRank 方法
  product.getRank();
  print(name);
}
```

抽象类和枚举

Dart 中抽象类的作用和用法与 Java 中的类似，都是通过 abstract 关键字将一个类声明为抽象类，然后在里面定义抽象方法、变量等，示例如下。

```
abstract class Dog {
  void doSomething(); // 定义抽象方法
}

class GoodDog extends Dog {
  void doSomething() {
    // 实现逻辑
  }
}
```

Dart 也支持枚举类型的定义，使用 enum 关键字可以声明枚举类型，作用和用法和其他面向对象语言中的类似，示例如下。

```
enum Color {
  red,
  green,
  blue
}
// 使用时直接调用
Color.blue
```

实现接口

Dart 支持多个接口实现，可以使用 implements 关键字。实现接口便于类的扩展、复用和解耦，从而提升开发效率，示例如下。

```
class Teacher {
  final _name;

  Teacher(this._name);

  String teach(who) => 'Hello';
}

class Student {
  Student();

  String study() => 'World!';
}
// 实现 Teacher 类和 Student 类接口
class ImlClass implements Teacher, Student {
  // ...
}
```

2.4.2　Dart 中的方法

方法也就是函数，Dart 中方法的定义和作用与 Java 等面向对象语言中的基本一致，大家可以对比着学习。Dart 中的方法用于实现具体逻辑和业务封装，我们可以通过传参来实现相关功能。接下来我们来看一下具体方法的实现。

构造方法

编写了类代码之后，有时需要传递一些参数来初始化这个类，这时可以定义构造方法进行传参。注意构造方法需要和类同名，示例如下。

```
class Position {
  num x;
  num y;

  Position(num x, num y) {
    this.x = x;
    this.y = y;
  }
}

// 也可以简化构造方法定义
class Point {
  num x;
  num y;

  Point(this.x, this.y);
}

void main() {
  var point = Point(2, 6);
  print(point.x);
}
```

静态方法

Dart 支持静态方法的定义和使用，使用静态方法时无须创建对象，使用时直接采用“类名.函数名”的形式进行调用即可。对于一些经常使用的方法，我们可以通过 static 关键字将其定义成静态方法，示例如下。

```
class Position {
  // 用 static 关键字定义静态方法
  static num getLongPosition() {
    return 20;
```

```
  }
}

void main(){
    // 使用时直接调用
    Position.getLongPosition();
}
```

可选参数方法

Dart 中另一个颇具特色的特性就是具有可选参数方法，我们可以不将参数全部传递到方法中，而只选择部分参数进行传递，示例如下。

```
// 普通方法
void Phone(int imei, String company, int money, String brand, int size) {
  // 具体逻辑
}
// 可选参数方法
void Phone2({int imei, String company, int money, String brand, int size}) {
  // 具体逻辑
}

void main() {
  // 普通方法，参数必须全部传递
  Phone(65408965, 'Apple', 3000, 'Apple', 5);
  // 可选参数方法，参数可以部分传递
  Phone2(imei: 6786876, company: 'Apple');
}
```

getter 方法和 setter 方法

getter 方法和 setter 方法分别用于取值和赋值操作，Dart 中无须单独声明 getter 方法和 setter 方法，直接调用即可，示例如下。

```
class Student {
  int id;
  int age;
  String name;
```

```
  int getRank() {
    return 20;
  }
}

void main() {
  var student = Student();
  // 赋值，等同于 setter 方法
  student.name = 'Tom';
  // 取值，等同于 getter 方法
  print(student.name);
}
```

匿名函数

Dart 中的匿名函数和 Java 中的匿名函数的作用和用法基本一致，即可以在传递的参数里直接实现方法，示例如下。

```
void _incrementCounter() {
  // 可以直接在 setState 中实现方法
  setState({
    _counter++;
  });
}
```

泛型类型

泛型类型一般用于进行多样化类型扩展。Java 中的泛型类型使用 T 来表示，Dart 中同样可以使用 T 来表示泛型类型，示例如下。

```
abstract class Dog<T> {
  T getDogByName(String name);
  setDogByname(String name, T value);
}

// 可以限制泛型类型继承自什么类
class Foo<T extends SomeBaseClass> {...}

class Extender extends SomeBaseClass {...}
```

```
void main() {
  var someBaseClassFoo = new Foo<SomeBaseClass>();
  var extenderFoo = new Foo<Extender>();
  var foo = new Foo();
}
```

2.5　Dart 的异步操作与导入类

Dart 是支持异步操作的，例如在网络请求等情况下，我们就需要用到异步操作。Dart 中使用 async 和 await 关键字来标识异步操作，不过需要注意，async 和 await 要成对出现，示例如下。

```
// 定义一个异步方法
getVersion() async {
  return '2.0.0';
}

// 定义一个异步方法
task() async {
  // 用 await 关键字等待获取这个异步方法的结果
  var version = await getVersion();
  print(version);
}

void main() {
  // 调用
  task();
}
```

除了 async 和 await，Dart 里也提供了 Future 对象，同样可以进行异步操作。我们可以让异步函数返回 Future 对象，然后通过这个 Future 对象来进一步获取结果，执行下一步逻辑，这有点类似于 RxJava 这种函数式编程方式。这种方式功能非常强大，使用起来也非常方便，示例如下。

```
// 用 Future 包装定义一个异步方法
Future<String> getVersion() async {
  return '2.0.0';
}
```

```
// 异步获取值
task() async {
  // 用 then 方法等待获取异步操作的结果
  getVersion().then((version) {
    print(version);
  });
}

void main() {
  // 调用
  task();
}
```

在 await 表达式中，返回值通常是一个 Future 对象，如果返回值不是 Future 对象，则 Dart 会自动将该值放到 Future 中返回。

在 Dart 中导入类也比较简单，通常使用 import 关键字来实现，示例如下。

```
import 'dart:io';
import 'package:mylib/mylib.dart';
import 'package:utils/utils.dart';
// 如果两个导入的类中有重名的，可以使用 as 关键字
import 'package:utils2/utils2.dart' as utils2;

// 也可以只导入一小部分，例如只导入 foo 类
import 'package:lib1/lib1.dart' show foo;

// 除了 foo 类，也导入其他类
import 'package:lib2/lib2.dart' hide foo;

// 延迟导入，减少 App 启动时间，优化性能
import 'package:deferred/hello.dart' deferred as hello;
// 可以多次调用 loadLibrary()函数，而只导入一次
greet() async {
  await hello.loadLibrary();
  hello.printGreeting();
}
```

如果我们想自己创建一个类并希望它被别人引用，这时可以用 library 来声明类，示例如下。

```
// 声明，名字为 abc
```

```
library abc;
// 导入需要用到的相关类
import 'dart:html';
```

如果想声明某个变量、常量、方法，使其不能被外部调用，只需要在其名字前加上下画线（_）前缀即可。Dart 的基础语法知识基本介绍完毕，当然这其中还有很多细节，感兴趣的读者可以自行深入研究。

第3章 Flutter开发入门

相信通过前面两章的介绍，大家对Flutter已经有了一个较为全面的认识。那么本章开始就进入Flutter开发环节，涉及开发环境的搭建、项目结构的分析、配置文件详解、组件分类，以及创建Flutter应用。希望大家通过本章的学习可以打下扎实的Flutter开发基础。

3.1 开发环境搭建

支持Flutter开发的IDE有很多，比如Android Studio、Visual Studio Code（VSCode）、InteIIiJ IDEA、Atom、Komodo等。本节将讲解Android Studio和VSCode。Android Studio和VSCode各有优缺点：Android Studio是Google官方推荐的开发工具，自带很多强大功能及模拟器；VSCode则更加轻量级、占用资源少。

关于模拟器，这里推荐使用Android官方模拟器，即Android Studio里自带的模拟器。我们可以单独启动模拟器，无须从Android Studio中启动，当然也可以用真机运行调试。

接下来，我们开始搭建Flutter开发环境。说明一下，本书是在Windows环境下搭建开发环境的。

3.1.1 Android Studio开发环境的搭建

本节将介绍在官方推荐的开发工具Android Studio下搭建开发环境的流程。

Flutter SDK 的下载与环境变量配置

截止到本书写作之时，Flutter SDK 的官方最新推荐版本是 1.17 版本，我们可以在官方网站下载最新版本的 SDK，如图 3-1 所示。

图 3-1　Flutter SDK 下载

Flutter SDK 的官方 GitHub 上主要有 dev、beta 和 stable 三个分支，正式开发时可以下载 stable 稳定版本，命令如下。

```
git clone -b stable https://github.com/flutter/flutter.git
```

下载完 Flutter SDK，我们可以把它解压并存入指定文件夹，然后配置 SDK 环境变量。将 Flutter 的 bin 目录加入环境变量，命令如下。这样一来，环境变量将配置完成。

```
[你的 Flutter 文件夹路径]\flutter\bin
```

接下来在命令提示符窗口输入如下命令，检查是否配置成功。

```
flutter doctor
```

如果有相关的错误提示，根据提示进行修复即可。每次运行这个命令都会检查是否缺失了必要的依赖。如果配置全都正确，会出现如图 3-2 所示的 Flutter 配置检查界面，主要的检查信息有 Flutter、Android toolchain、Connected device、开发工具 IDE。

```
命令提示符
Microsoft Windows [版本 10.0.17134.829]
(c) 2019 Microsoft Corporation。保留所有权利。

C:\Users\Home>flutter doctor
Doctor summary (to see all details, run flutter doctor -v):
[√] Flutter (Channel stable, v1.17.1, on Microsoft Windows [Version 10.0.17134.829], locale zh-CN)

[√] Android toolchain - develop for Android devices (Android SDK version 28.0.3)
[√] Android Studio (version 3.6)
[√] IntelliJ IDEA Community Edition (version 2018.3)
[!] Connected device
    ! No devices available

! Doctor found issues in 1 category.

C:\Users\Home>_
```

图 3-2　Flutter 配置检查界面

Flutter 和 Dart 插件安装与项目新建

Android Studio 的安装很简单，下载相应的安装包后按步骤操作即可，这里不做详细讲解。Android Studio 安装完成后，我们启动它来安装 Flutter 和 Dart 插件。

这里使用的是 Android Studio 3.3.1 版本，大家也可以使用 3.4 及以上版本。我们打开 Android Studio，选择【File】，然后选择【Settings】，在【Plugins】下搜索 Flutter 和 Dart 插件进行安装，如图 3-3 所示。

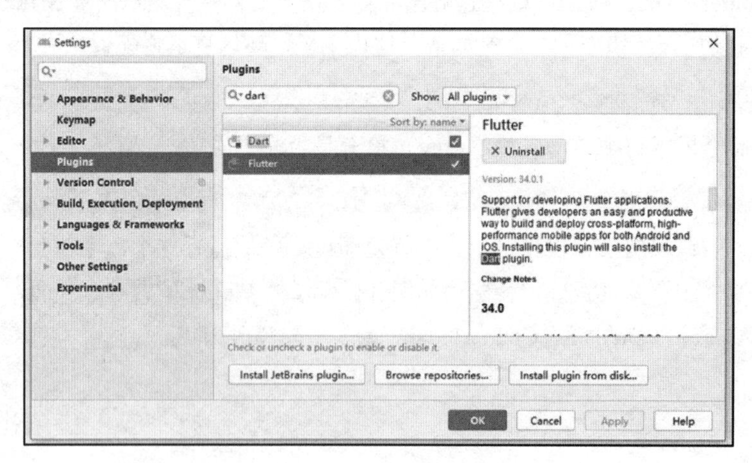

图 3-3　安装 Flutter 和 Dart 插件

安装好插件后重启 Android Studio，这样一来，基本环境就搭建好了。接下来我们选择【File】，然后依次选择【New】和【New Flutter Project】，这样就可以新建 Flutter 项目了，如图 3-4 所示。

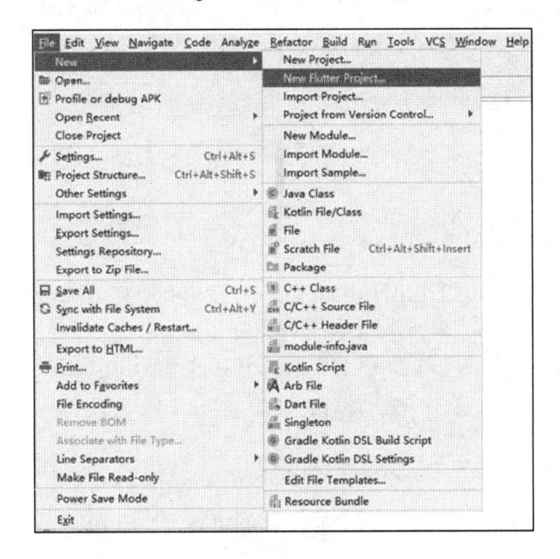

图 3-4　新建 Flutter 项目

经过以上步骤，会弹出新建 Flutter 项目的选择界面，此时选择【Flutter Application】并选择【Next】，完成项目选择，进入配置项目界面。根据提示连续选择【Next】，完成项目配置、包名信息填充。设置好所有信息后，选择【Finish】就可以完成项目新建了。

Flutter 项目结构如图 3-5 所示。

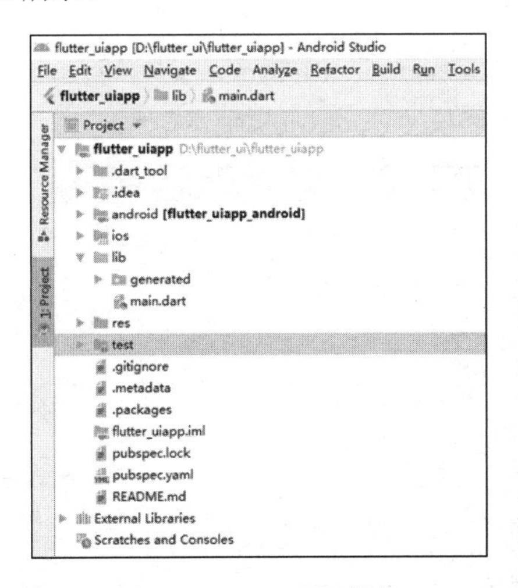

图 3-5　Flutter 项目结构

3.1.2 VSCode 开发环境的搭建

VSCode 相对于 Android Studio 而言是一个比较轻量级的开发工具，大家可以根据自己的习惯来选择开发工具。接下来我们介绍 VSCode 开发环境的搭建流程。

下载 Flutter SDK

Flutter SDK 由两部分构成，一部分是 Dart SDK，另一部分是 Flutter SDK。可以通过两种方式下载 Flutter SDK，一种是通过 GitHub 下载，另一种是直接下载 SDK 压缩包。

通过 GitHub 下载时，我们可以拉取官方 GitHub 上的 Flutter 分支。官方分支主要有 dev、beta 和 stable 三个，要想进行正式开发，建议下载 stable 稳定版本，命令如下。

```
git clone -b stable https://github.com/flutter/flutter.git
```

另一种方式是直接在官网下载 SDK 压缩包，前面已经讲过，这里就不重复介绍了。

环境变量配置

关于环境变量配置，VSCode 和 Android Studio 是一样的，这里不再赘述。

安装 VSCode 所需的插件

我们需要在 VSCode 的 Extensions 里搜索并安装 Dart 和 Flutter 的扩展插件，如图 3-6 所示。安装完成后，重启 VSCode 编辑器即可完成环境搭建。

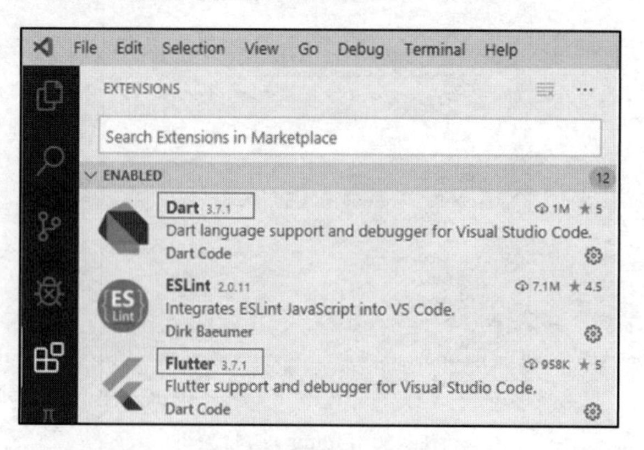

图 3-6　安装 Dart 和 Flutter 的扩展插件

新建 Flutter 项目

新建 Flutter 项目时，我们可以直接打开命令面板，或通过组合键 Ctrl+Shift+P 打开命令面板，找到【Flutter：New Project】，如图 3-7 所示。

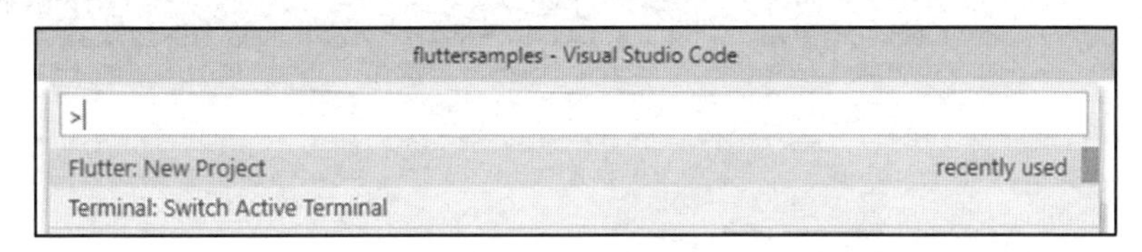

图 3-7 新建 Flutter 项目

然后完善内容，输入项目名称，选择项目存储位置，这样就完成了 Flutter 项目的新建。新建完毕的 Flutter 项目结构如图 3-8 所示。

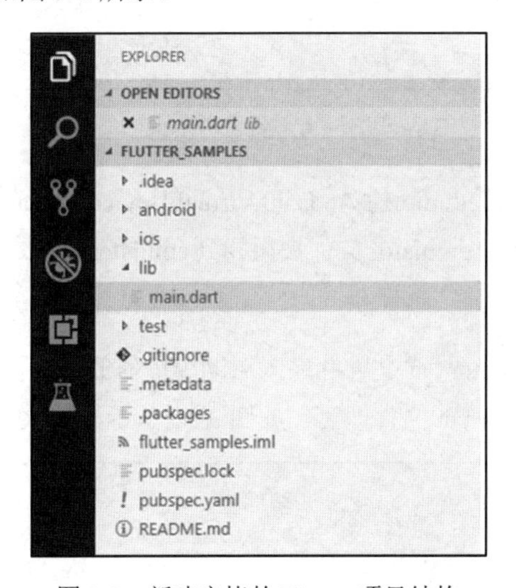

图 3-8 新建完毕的 Flutter 项目结构

其中，与 Android 相关的修改和配置项位于 android 目录下，其结构与 Android 应用的项目结构一致；与 iOS 相关的修改和配置项位于 ios 目录下，结构与 iOS 应用的项目结构一致。最重要的 Flutter 代码文件存于 lib 目录下，类文件以.dart 结尾。

3.1.3 模拟器的新建与调试

项目新建完毕，接下来就要进行编译，并将 Flutter 项目运行到真机或模拟器上了。首先我们介绍如何在 Android Studio 中新建模拟器。

选择 Android Studio 顶部的【Android Virtual Device Manager】，如图 3-9 所示，这样即可完成模拟器的新建。

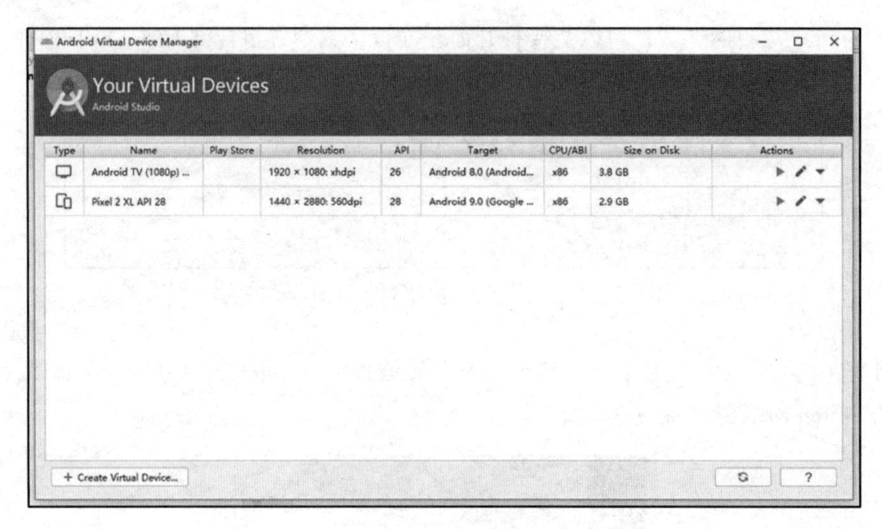

图 3-9　新建模拟器

如果我们不想从 Android Studio 的 Android Virtual Device Manager 里新建和管理模拟器，也可以在 Android SDK 目录下的 emulator 文件夹里找到 emulator.exe 文件，从而启动 Android Virtual Device Manager。

我们可以建立一个 emulator.bat 文件并写入启动模拟器的命令，这样一来，每次启动模拟器时只要直接运行这个 emulator.bat 文件即可，如图 3-10 所示。

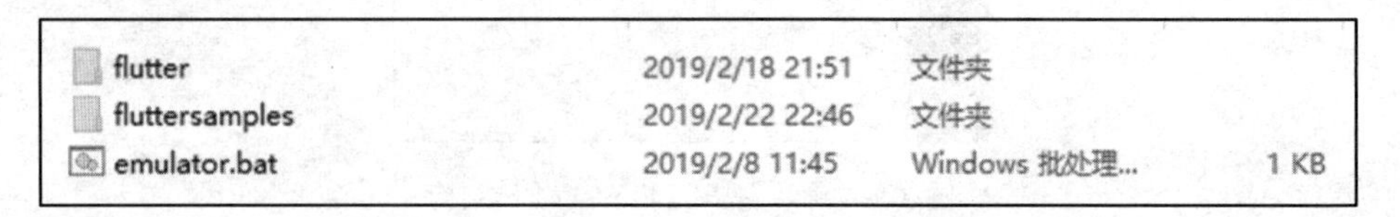

图 3-10　bat 文件

用鼠标左键双击 emulator.bat 文件，运行模拟器，此时界面如图 3-11 所示。

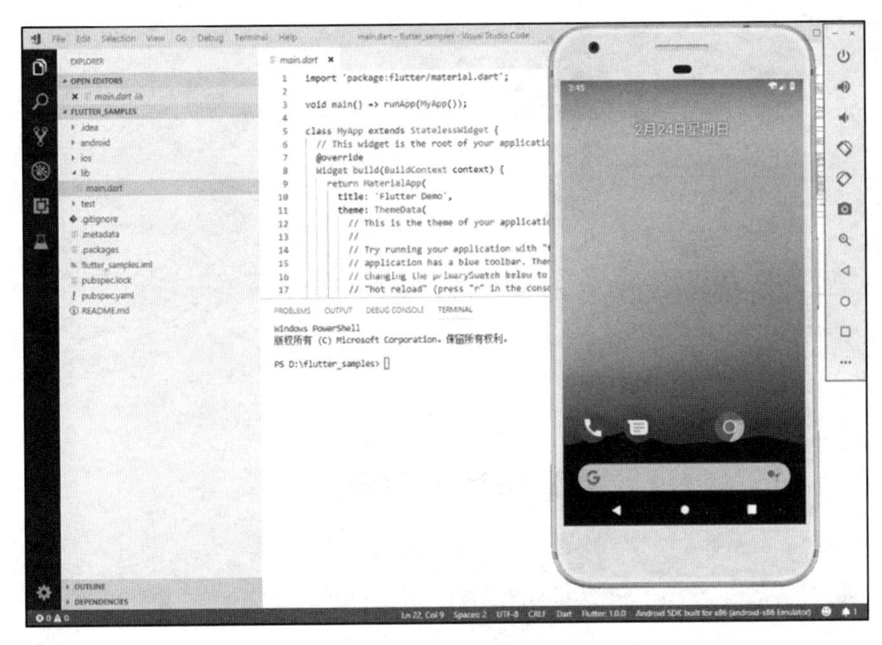

图 3-11　运行模拟器界面

接着在项目所在目录运行 flutter run 命令，启动模拟器日志，将 Flutter 项目在模拟器上编译运行，如图 3-12 所示。

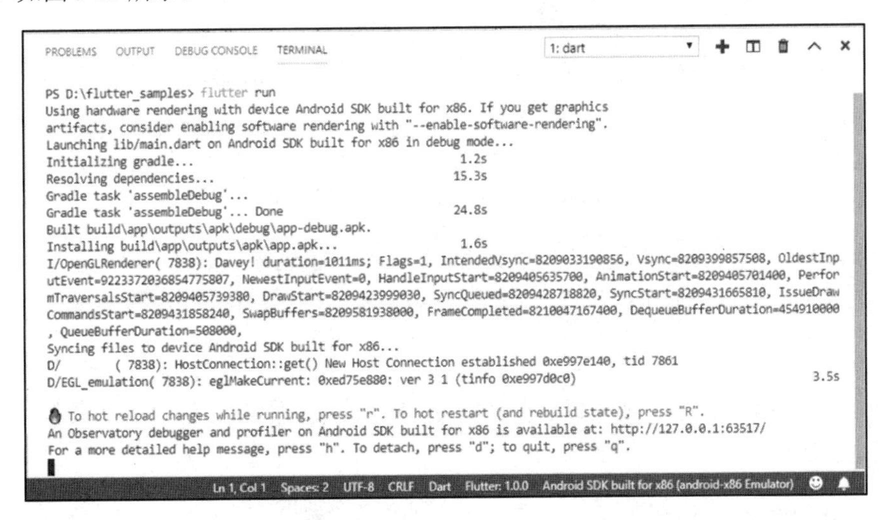

图 3-12　启动模拟器日志

运行效果如图 3-13 所示。

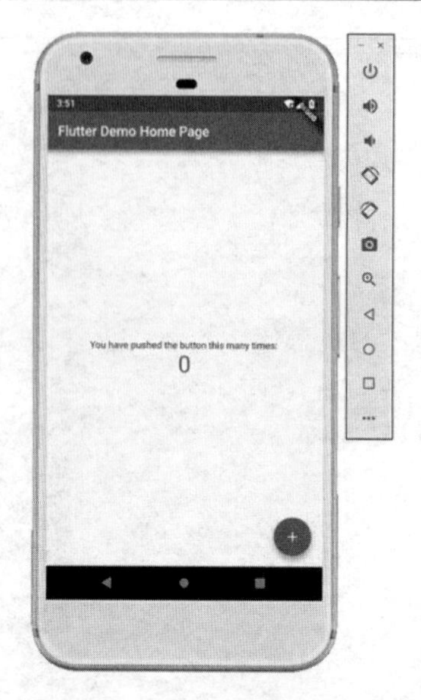

图 3-13　Flutter 项目在模拟器上的运行效果

在开发环境下可以进行项目的热重载（Hot Reload）和热重启（Hot Restart），操作非常方便。如图 3-14 所示，闪电形状的图标为热重载按钮，它右侧的环形带闪电的图标为热重启按钮。

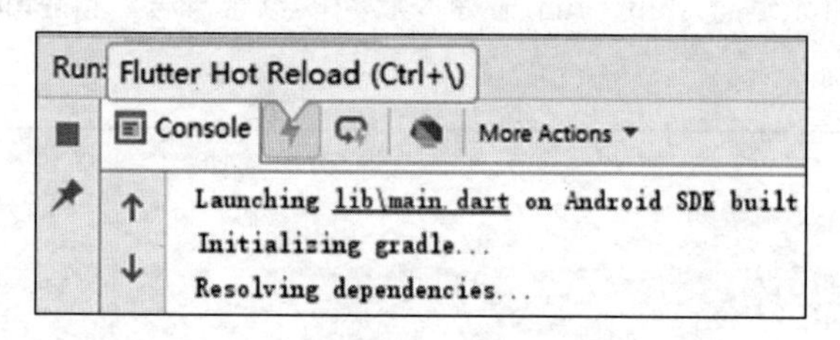

图 3-14　热重载按钮和热重启按钮

Flutter 项目运行成功后，只要不退出应用界面，就可以进行热重载。输入 r 会热重载当前页面，输入 R 将进行整个应用的热重启，输入 h 弹出帮助信息，输入 d 解除关联，输入 q 退出应用调试。

如果遇到有多个模拟器，或模拟器和真机同时存在的情况，可以通过“-d 参数+设备 ID”的形式指定要运行的设备，命令如下。

```
flutter run -d emulator-5556
```

真机设备运行调试和模拟器运行调试的过程基本一致，将手机和电脑通过 USB 连接，手机开启开发人员选项和 USB 调试，最后运行 flutter run 命令即可。下面列出几个常用的命令，大家要牢记这些命令，并在开发过程中熟练使用它们。

```
flutter build apk;              // 打包 Android 应用
flutter build apk -release;
flutter install;                // 安装应用
flutter build ios;              // 打包 iOS 应用
flutter build ios -release;
flutter clean;                  // 清理并重新编译项目
flutter upgrade;                // 升级 Flutter SDK 和依赖包
flutter channel;                // 查看 Flutter 官方分支列表和当前项目使用的 Flutter 分支
flutter channel <分支名>;       // 切换分支
```

关于开发环境搭建和模拟器配置的内容就介绍这么多，建议大家使用稳定版 Flutter SDK，尝试新建一个项目并将它运行到手机或模拟器上，做到熟练上手。

3.2　项目结构分析

了解项目结构对于深入学习 flutter 有很重要的作用。在新建了 Flutter 项目后，需要进一步了解 Flutter 的项目结构、配置文件的作用等。由于 Flutter 应用是跨平台应用，所以一般会包括 android 和 ios 两个目录。新建项目后，一般会看到如图 3-15 所示的 Flutter 默认项目结构。

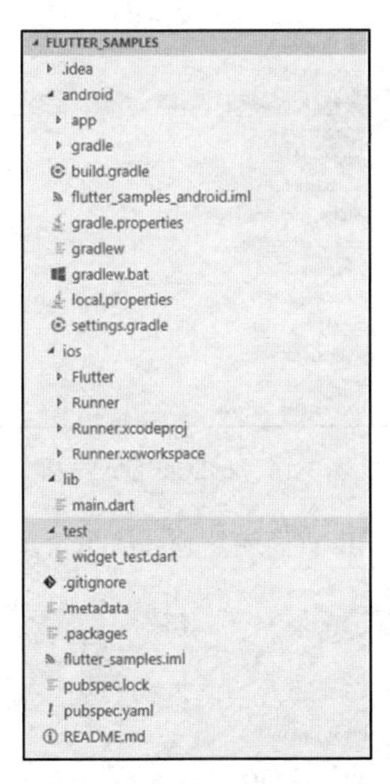

图 3-15　Flutter 默认项目结构

可以看到，其中主要包含了 android、ios、lib、test 几个目录，以及 pubspec.yaml 配置文件等。下面我们逐一介绍这几个目录的作用。

1. android 目录

android 目录显示了一个完整的 Android 项目包含的内容，编译后，Flutter 代码会整合到这个 Android 项目代码中，形成一个 Flutter Android 应用。通过 Android Studio 打开并预览 android 目录，其结构如图 3-16 所示。

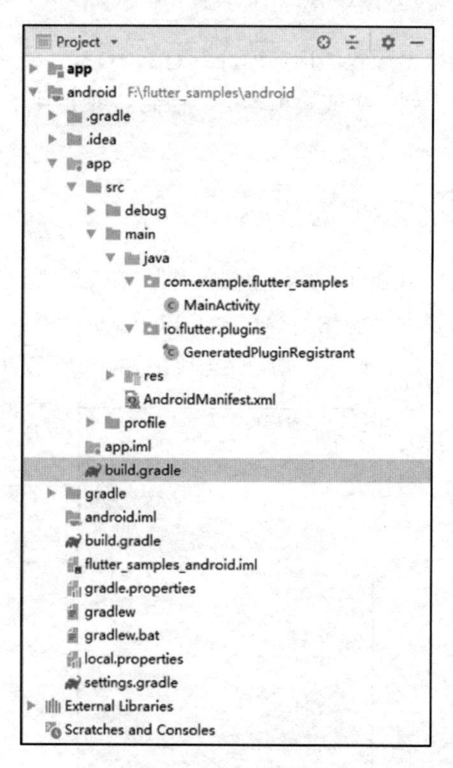

图 3-16　android 目录结构

Android 项目默认启动的 Activity 页面就是 MainActivity，里面的内容如下。

```
package com.example.flutter_samples;

import android.os.Bundle;
import io.flutter.app.FlutterActivity;
import io.flutter.plugins.GeneratedPluginRegistrant;
```

```
public class MainActivity extends FlutterActivity {
  @Override
  protected void onCreate(Bundle savedInstanceState) {
    super.onCreate(savedInstanceState);
    GeneratedPluginRegistrant.registerWith(this);
  }
}
```

可以看到，里面几乎没有什么代码逻辑，只是继承了 FlutterActivity，然后将 Flutter 相关插件注册进来。后续如果我们要编写插件，需要在这里写入一些回调插件逻辑。

如果需要修改 Android 应用的包名、版本号、版本名称，实现方式与修改 Android 项目是一样的，都是在 app 目录下的 build.gradle 里进行的，代码如下。

```
android {
    compileSdkVersion 28
    ...
    defaultConfig {
        // 包名
        applicationId "com.example.flutter_samples"
        minSdkVersion 16
        targetSdkVersion 28
        // 版本号
        versionCode 1
        // 版本名称
        versionName "1.0"
        testInstrumentationRunner "android.support.test.runner.AndroidJUnitRunner"
    }
...
```

2. ios 目录

ios 目录和 android 目录一样，其存放的是 iOS 项目文件，涵盖了一个完整的 iOS 项目。我们可以使用 XCode 打开 iOS 项目，并执行编译、修改、开发、调试等操作，例如完成 iOS 平台图标的修改、版本号的修改等，这里不再赘述。

3. lib 目录

lib 目录中存放了核心的 Flutter 代码逻辑：Dart 语言的类源代码文件（例如 main.dart）。最终 lib 下的源代码会被编译到 Android 和 iOS 平台并进行渲染。main.dart 类的名字不可以修改，必须放置在 lib 根目录下，它是项目的入口文件，即入口类。在后面的章节中，我们会结合注释详细介绍 main.dart，这里先给大家留一个悬念。

在实际开发中，我们还需要进行包划分，如图 3-17 所示。我们可以在 lib 里建立相关的文件夹，按照项目功能、类的类型和功能进行包划分。

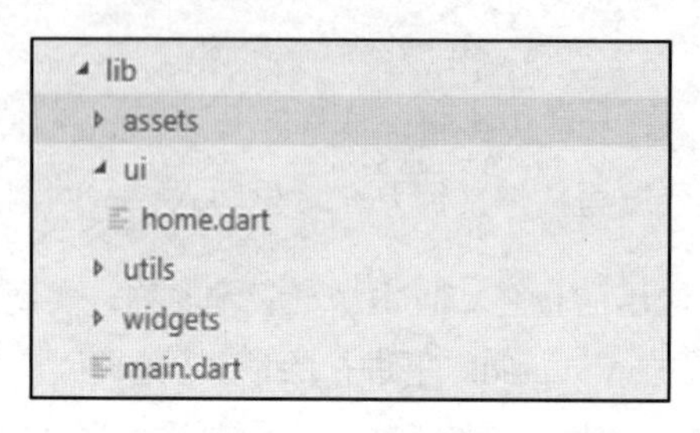

图 3-17　包划分

4．test 目录

test 目录里主要存放编写的测试用例，如测试 UI 组件、数据等。test 目录里面默认有一个 widget_test.dart 文件，用于编写测试用例。我们可以运行、修改、扩展其中的测试用例，实现 Flutter 代码测试。

关于 Flutter 项目结构和目录作用的内容介绍得差不多了，了解和掌握 Flutter 项目结构和目录作用对于项目开发和应用架构都非常有帮助。

3.3　配置文件详解

很多项目都有自己的构建系统或配置文件， Flutter 同样有自己的项目配置文件，用于配置项目相关参数。主要的文件是 pubspec.yaml，我们可以在这里配置和引用第三方插件库、添加 assets 图片资源、添加 fonts 字体资源、添加音视频资源路径等。整个文件是参照 YAML 语法规范进行定义的。以下是 pubspec.yaml 文件的内容和结构。

```
// 项目名称用英文，类似于 Android 中的包名，如果修改它则整个项目的引入路径都要修改
// 确定了就不要再修改
name: flutter_samples
// 项目描述
description: A new Flutter project.

// 版本号，覆盖对应的 Android 和 iOS 应用版本号
```

```
// +前对应 Android 的 versionCode，+后对应 Android 的 versionName
// +前对应 iOS 的 CFBundleVersion，+后对应 iOS 的 CFBundleShortVersionString
version: 1.0.0+1

// 项目的编译环境为 Dart SDK，版本号在 2.1.0 和 3.0.0 之间
environment:
  sdk: ">=2.1.0 <3.0.0"

// 项目的依赖插件库
// Flutter 插件库在 Dart Pub 里查找、引用
dependencies:
  flutter:
    sdk: flutter

// 在此处引入插件库
  cupertino_icons: ^0.1.2
  flutter_webview_plugin: ^0.3.1

dev_dependencies:
  flutter_test:
    sdk: flutter

// 相关配置
flutter:
// 是否使用 material 图标，建议设为 true
  uses-material-design: true

  // 配置项目文件里的图片路径
  // 如果需要使用项目目录内附带的图片、音频、视频等资源，必须在这里配置
  assets:
    - images/a_dot_burr.jpeg
    - images/a_dot_ham.jpeg

  // 字体文件资源配置
  fonts:
    - family: Schyler
      fonts:
        - asset: fonts/Schyler-Regular.ttf
        - asset: fonts/Schyler-Italic.ttf
```

```
        style: italic
    - family: Trajan Pro
      fonts:
        - asset: fonts/TrajanPro.ttf
        - asset: fonts/TrajanPro_Bold.ttf
          weight: 700
```

再多说一句，如果我们在配置文件里引入一个插件库（如 shared_preferences），则再次用 Android Studio 打开 Android 项目时，可以看到目录的依赖项里增加了 shared_preferences，如图 3-18 所示。同时，Android 项目主工程 GeneratedPluginRegistrant 类里也会增加新内容。

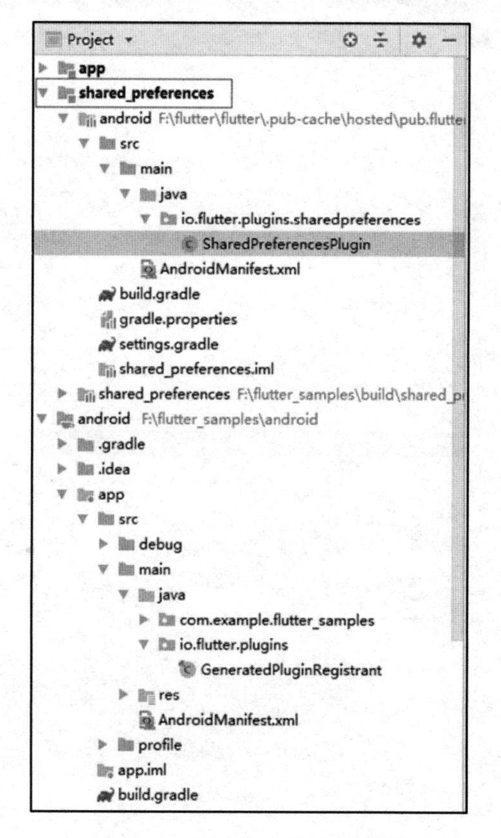

图 3-18　增加依赖项的目录结构

除了 pubspec.yaml 这个核心配置文件，还有 pubspec.lock、.packages、.metadata 等文件。

- pubspec.lock：指明项目引用的依赖库的具体版本号等信息，如果某个配置文件丢失，可以通过这个文件重新下载和恢复依赖库，属于自动生成的文件。

- .packages：存放项目依赖库在本机上的绝对路径，属于自动生成的文件，如果项目出错或无法找到某个库，可以把这个文件删除，重新配置。
- .metadata：记录项目的属性信息，如开发时使用的 Flutter SDK 分支信息，项目属性是什么，属于自动生成的文件，无须修改、删除。

关于 Flutter 配置文件的内容就介绍到这里，大家可以根据以上内容对配置文件进行相关操作，通过实践加深理解。

3.4　Flutter 组件化

Flutter 的面向对象编程方式借鉴了 React 的组件化编程思维。Flutter 中的所有类都可以看成组件，大部分类都继承自 Widget 类。本节主要介绍 Flutter 整体架构层级和 Flutter 组件分类，同时会覆盖一些其他关于组件的知识点。

3.4.1　架构层级

前面的章节中介绍过 Flutter 的架构，这里再来回顾一下。Flutter 架构整体分为两层：Engine 层和 Framework 层，如图 3-19 所示。我们最常接触和调用 API 的层级就是 Framework 层，所以本节主要分析 Framework 层。

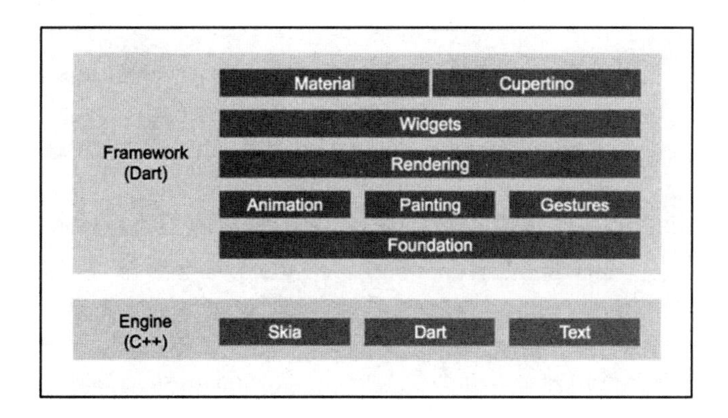

图 3-19　Flutter 架构

Framework 层中的底层是 Foundation 层，核心是 foundation 模块，其中包含绑定和注解等基础功能类。从 foundation 模块中包含的类可以看出，foundation 模块主要负责一些基础操作，如注解、断言、绑定、更新通知、编码等。

Foundation 层的上一层有三个部分，核心分别为 animation、painting、gestures 模块。

- animation 模块主要用于放置一些与动画相关的 API 组件，如 animation.dart、curves.dart、animation_controller.dart、tween.dart 等都非常常用。
- painting 模块主要用于存放边框绘制、裁剪图像处理、插值器等与绘制、装饰相关的组件，如 alignment.dart、box_decoration.dart、decoration.dart、edge_insets.dart、image_provider.dart、colors.dart、text_style.dart、text_painter.dart 等。
- gestures 模块，顾名思义，里面存放了与手势处理相关的组件，基本涵盖了拖曳、长按、触摸、放大等各种功能组件，如 drag.dart、events.dart、long_press.dart、multidrag.dart、multitap.dart、tag.dart、scale.dart。

继续往上一层看——Rendering 层。其核心 rendering 模块可以看作渲染库，是组件的父类基础库，里面涵盖了一些渲染树，如 flex.dart、flow.dart、image.dart、sliver.dart、stack.dart、table.dart、view.dart 等。

再往上看，Rendering 层的上一层是 Widgets 层，Widgets 层依赖于 Rendering 层进行构建，里面的功能组件非常多，因为 Flutter 里所有的类都可以看作组件，常用的有 app.dart、container.dart、bottom_navigation_bar_item.dart、editable_text.dart、form.dart、gesture_detector.dart、icon.dart、image.dart、navigator.dart、page_view.dart、routes.dart、scroll_view.dart、sliver.dart、spacer.dart、text.dart 等。

最后，我们来看一下 Framework 层的顶层，顶层对应 material 和 cupertino 模块。这两个模块分别对应 Android 平台风格的 Material 组件和 iOS 平台风格的扁平化组件。我们可以直接使用这两套风格的组件去构建符合 Android 和 iOS 平台风格的应用。

- material 模块里的组件非常多，大部分是我们常用的，如 app_bar.dart、button.dart、card.dart、bottom_app_bart.dart、bottom_navigation_bar.dart、colors.dart、data_table.dart、dialog.dart、icons.dart、snack_bar.dart、scaffold.dart、text_field.dart、theme.dart 等。
- cupertino 模块里的组件目前并不多，常用的如 action_sheet.dart、bottom_tab_bar.dart、button.dart、colors.dart、dialog.dart、icons.dart、nav_bar.dart、page_scaffold.dart、tab_scaffold.dart、tab_view.dart、text_field.dart、theme.dart 等。

3.4.2 组件分类

介绍了 Flutter 整体架构层次后，我们着重看一下 Flutter 的核心，即组件分类。图 3-20 是 Flutter 官方给出的组件分类结构图。

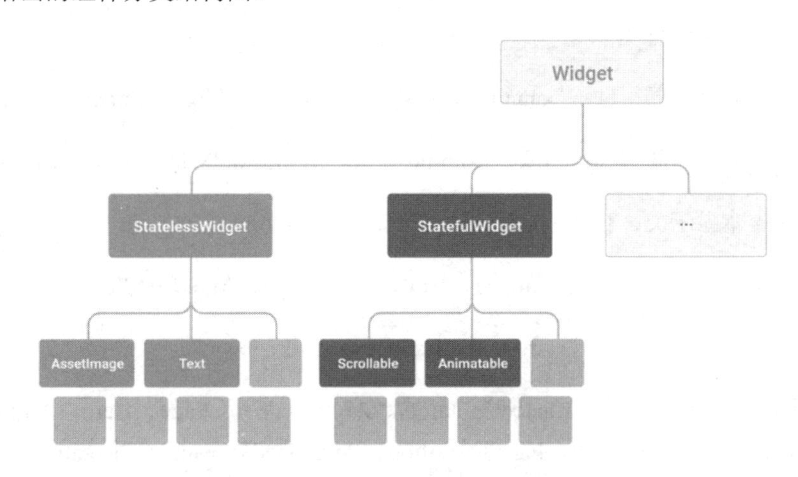

图 3-20 Flutter 官方给出的组件分类结构图

可以看出，组件主要分为 StatelessWidget 和 StatefulWidget 两类，分别表示无状态组件和有状态组件。无状态组件主要是指那些无须更新页面显示状态的、无可变状态维护功能的组件，如 AssetImage、Text 等。有状态组件主要是指那些可以自己维护状态、更新渲染内容的组件，如 Scrollable、Animatable 等。

Flutter 官方将组件分为以下几类。

- 基础组件（WidgetBasics）：Container、Row、Column、Image、Text、Icon、RaisedButton、Scaffold、Appbar、FlutterLogo、Placeholder。
- Material 组件类
- App 结构和导航类：Scaffold、Appbar、BottomNavigationBar、TabBar 等。
- 按钮类：RaisedButton、FloatingActionButton、FlatButton、IconButto 等。
- 输入和选择类：TextField、Checkbox、Raido、Switch、Slider 等。
- 对话框和控制面板类：SimpleDialog、AlertDialog、BottomSheet 等。
- 信息显示类：Image、Icon、Chip、Tooltip、DataTable 等。

- 布局类：ListTile、Stepper、Divider。
- Cupertino 组件类：CupertinoActionSheet、CupertinoAlertDialog、CupertinoButton、CupertinoDatePicker、CupertinoDialog、CupertinoDialogAction 等。
- Layout 布局组件类
- 单个子元素布局组件：Container、Padding、Center、Align、FittedBox 等。
- 多个子元素布局组件：Row、Column、Stack、IndexedStack、GridView 等。
- Text 文本显示类：Text、RichText、DefaultTextStyle。
- Assets、图片、Icons 类：Image、Icon、RawImage、AssetBundle。
- Input 输入类：Form、FormField、RawKeyboardListener。
- 动画和 Motion 类：AnimatedContainer、AnimatedCrossFade、Hero、AnimatedBuilder、DecoratedBoxTransition、FadeTransition、PositionedTransition、RotationTransition 等。
- 交互模型类
- 触摸交互：Draggable、LongPressDraggable、GestureDetector、DragTarget 等。
- 路由导航：Hero、Navigator。
- 样式类：Padding、Theme、MediaQuery。
- 绘制和效果类：Transform、Opacity、DecoratedBox、RotatedBox、ClipOval、ClipPath、ClipRect、CustomPaint、BackdropFilter 等。
- Async 异步模型类：FutureBuilder、StreamBuilder。
- 滚动类：GridView、ListView、NestedScrollView、SingleChildScrollView 等。
- 辅助功能类：Semantics、MergeSemantics、ExcludeSemantics。

本节主要介绍 Flutter 的组件分类，为后续的 Flutter 开发奠定基础，熟悉组件的分类会更利于我们掌握组件的用法。

3.5　创建 Flutter 应用

本节我们将尝试编写 Flutter 应用，感受 Flutter 的语法特点和运行效率。首先我们将创建一个默认应用，接着尝试编写自己的应用，一起开始吧！

3.5.1　创建默认应用

本节用到的开发工具是 VSCode，当然也可以使用 Android Studio。我们要编写的应用整体功能很简单，主要涉及的内容有控件点击事件、setState(() {...}) 更新、Text 组件显示等。

接下来通过代码结合注释的方式来看一下这个应用的 main.dart 的实现，具体如下。

```
import 'package:flutter/material.dart';

// main.dart 为应用入口类，里面的 void main()为入口函数
// void main(){
//   return runApp(MyApp());
// }
void main() => runApp(MyApp());

class MyApp extends StatelessWidget {
  // 这个组件是应用的根布局，类似页面容器
  // 构建页面
  @override
  Widget build(BuildContext context) {
    // 入口页面使用 MaterialApp 页面脚手架构建
    // 可以快速构建页面
    // MaterialApp 脚手架默认自带顶部 ToolBar、路由、主题、国际化等配置
    return MaterialApp(
      title: 'Flutter Demo',
      theme: ThemeData(
        // 这里可以配置应用全局主题
        // 可以通过 flutter run 命令运行程序，会看到蓝色状态栏和标题栏
        // 通过 primarySwatch 属性来配置状态栏和标题栏颜色
        primarySwatch: Colors.green,
      ),
      // 设置页面启动组件
      home: MyHomePage(title: 'Flutter Demo Home Page'),
```

```
    );
  }
}

// 继承自 StatefulWidget，有状态组件
class MyHomePage extends StatefulWidget {
  // 有参构造方法，用来传值
  MyHomePage({Key key, this.title}) : super(key: key);

  final String title;

  // 重写创建状态
  @override
  _MyHomePageState createState() => _MyHomePageState();
}

  // 自定义创建状态管理，继承自 State<T>
 class _MyHomePageState extends State<MyHomePage> {
  // 声明变量临时存储次数
  int _counter = 0;
  // 定义方法来累加次数
  void _incrementCounter() {
    setState(() {
      // 刷新 UI 和绑定数据
      _counter++;
    });
  }

  @override
  Widget build(BuildContext context) {
    // 每次调用 setState 时都会调用 build 方法
    // 构建页面布局，这里使用了 Scaffold 脚手架
    // 包含 AppBar、body、bottomNavigationBar、floatingActionButton 等
    return Scaffold(
      appBar: AppBar(
        // 通过配置 AppBar 属性来控制显示效果，这里通过 title 来设置标题内容
        title: Text(widget.title),
      ),
      body: Center(
```

```
      // body 部分用 Center 来加载组件布局内容，子组件居中排列
      child: Column(
        // Column 是纵向列布局，子组件纵向排列
        mainAxisAlignment: MainAxisAlignment.center,
        children: <Widget>[
          // 子组件，Text 组件用来显示文字内容
          Text(
            'You have pushed the button this many times:',
          ),
          // 动态绑定数据
          Text(
            '$_counter',
            style: Theme.of(context).textTheme.display1,
          ),
        ],
      ),
    ),
    // 浮动+按钮
    floatingActionButton: FloatingActionButton(
      // 设置点击事件，执行_incrementCounter 方法累加计数
      onPressed: _incrementCounter,
      // 设置长按提示信息
      tooltip: 'Increment',
      // 设置图标
      child: Icon(Icons.add),
    ),
  );
 }
}
```

执行以上代码，在通过 primarySwatch 属性来配置状态栏和标题栏的颜色后，我们会看到顶部状态栏、标题栏、按钮的主题色调变成绿色。

一般的入口文件通过 MaterialApp 脚手架构建，其他页面可以不使用。我们先来看一下 MaterialApp 脚手架构造方法都提供了哪些可配置的属性功能，具体如下。

```
const MaterialApp({
  Key key,
  this.navigatorKey,
  this.home,
```

```
    this.routes = const <String, WidgetBuilder>{},
    this.initialRoute,
    this.onGenerateRoute,
    this.onUnknownRoute,
    this.navigatorObservers = const <NavigatorObserver>[],
    this.builder,
    this.title = '',
    this.onGenerateTitle,
    this.color,
    this.theme,
    this.darkTheme,
    this.locale,
    this.localizationsDelegates,
    this.localeListResolutionCallback,
    this.localeResolutionCallback,
    this.supportedLocales = const <Locale>[Locale('en', 'US')],
    this.debugShowMaterialGrid = false,
    this.showPerformanceOverlay = false,
    this.checkerboardRasterCacheImages = false,
    this.checkerboardOffscreenLayers = false,
    this.showSemanticsDebugger = false,
    this.debugShowCheckedModeBanner = true,
})
```

可以看到，通过入口文件的 MaterialApp 配置可以实现应用的整体显示和各类常用功能。接下来，我们再来看一下 Scaffold 脚手架构造方法提供的可配置属性功能，具体如下。

```
const Scaffold({
    Key key,
    this.appBar,
    this.body,
    this.floatingActionButton,
    this.floatingActionButtonLocation,
    this.floatingActionButtonAnimator,
    this.persistentFooterButtons,
    this.drawer,
    this.endDrawer,
    this.bottomNavigationBar,
    this.bottomSheet,
    this.backgroundColor,
```

```
    this.resizeToAvoidBottomPadding,
    this.resizeToAvoidBottomInset,
    this.primary = true,
    this.drawerDragStartBehavior = DragStartBehavior.down,
})
```

同样，我们可以看到，通过 Scaffold 脚手架可以很快地搭建出一个个性化页面，基本可以实现页面上所需的大部分功能，达到预期显示效果。

3.5.2 创建自己的应用

有了前面的内容作为基础，接下来我们自己动手编写一个简单的页面，实现“显示一段文字和一张图片，按下按钮切换文字内容”的效果，具体代码如下。

```
import 'package:flutter/material.dart';

void main() {
  return runApp(ShowApp());
}

class ShowApp extends StatelessWidget {
  @override
  Widget build(BuildContext context) {
    return MaterialApp(
      title: 'Flutter Demo',
      theme: ThemeData(
        primarySwatch: Colors.green,
      ),
      home: ShowAppPage(),
    );
  }
}

class ShowAppPage extends StatefulWidget {

  @override
  _ShowAppPageState createState() {
    return _ShowAppPageState();
  }
}
```

```
class _ShowAppPageState extends State<ShowAppPage> {
  String title = '春天的脚步近了，我们应该更具青春朝气';
  bool change = false;

  void _changeTextContent() {
    setState(() {
      // 刷新 UI 和绑定数据
      title = change ? "这张图很好看，体现了春天的气息" : "春天的脚步近了，我们应该更具青春朝气";
      change = !change;
    });
  }

  @override
  Widget build(BuildContext context) {
    return Scaffold(
      appBar: AppBar(
        title: Text('春天的气息'),
      ),
      body: Center(
        child: Column(
          mainAxisAlignment: MainAxisAlignment.center,
          children: <Widget>[
            Padding(
              padding: EdgeInsets.all(10),
              child: Image.network(
                  'https://***.com/***.jpg'),
            ),
            // 动态绑定数据
            Padding(
              padding: EdgeInsets.all(10),
              child: Text(
                '$title',
                style: Theme.of(context).textTheme.title,
              ),
            ),

            RaisedButton(
              onPressed: _changeTextContent,
```

```
                child: Text('切换内容'),
              ),
            ],
          ),
        ),
      );
    }
}
```

通过 flutter run 命令将以上示例编译、运行到真机或模拟器上，运行效果如图 3-21 所示。

图 3-21　Flutter 示例效果

怎么样？效果是不是很好？而且构建这个页面相比于用其他语言进行构建要简单得多，运行非常流畅，体验非常好。

本节主要给大家讲解了用 Flutter 创建应用的方法。俗话说“熟能生巧”，我们不但要理解理论知识，也要时常动手实践，熟练掌握实践方法才能更好地进行深入研究和开发。

第4章 Flutter开发规范

养成良好的开发习惯不但有利于提升自己的开发效率，也能让其他人更好理解自己编写的代码。Flutter的部分开发规范和约束与其他编程语言有一些区别，所以本章将着重讲解Flutter开发规范。主要内容包括：Flutter项目结构规范、Flutter命名规范、Flutter代码格式规范、Flutter注释规范，以及Flutter代码使用规范。

4.1 项目结构规范

前面我们讲过Flutter的项目结构，默认新建项目后，项目结构完全遵循官方标准项目结构要求。开发时也要按照这个标准要求设置目录结构，不能更改。

如果我们需要新建项目内的资源文件目录，并引入一些打包进去的图片文件、音频文件、视频文件、字体文件等，可以参照官方例子，在项目根目录下新建assets目录，并把项目资源文件存放在其中。如果有需要，还可以在assets目录里再次分类，建立audios、fonts、images、videos等资源目录。assets目录结构如图4-1所示。

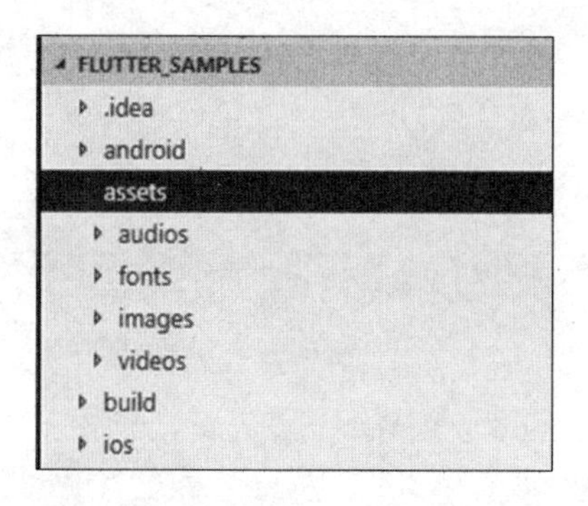

图4-1 assets目录结构

另外，我们还可以直接在项目根目录下创建多个类别资源文件夹，如创建 fonts 文件夹，用于存放字体文件资源，与用于存放图片、图标资源的 assets 文件夹并列，大致情况如图 4-2 所示。

电脑 › 文档 (F:) › flutter_samples ›

名称	修改日期	类型	大小
.idea	2019/3/18 22:10	文件夹	
android	2019/3/19 0:31	文件夹	
assets	2019/3/30 18:01	文件夹	
build	2019/3/19 0:31	文件夹	
fonts	2019/3/30 18:20	文件夹	
ios	2019/3/18 22:10	文件夹	
lib	2019/3/30 18:01	文件夹	
test	2019/3/18 22:10	文件夹	
.flutter-plugins	2019/3/30 18:18	FLUTTER-PLUGI...	1 KB
.gitignore	2019/3/18 22:10	文本文档	2 KB
.metadata	2019/3/18 22:10	METADATA 文件	1 KB
.packages	2019/3/30 18:18	PACKAGES 文件	3 KB
flutter_samples.iml	2019/3/18 22:10	IML 文件	1 KB
pubspec.lock	2019/3/30 18:18	LOCK 文件	4 KB
pubspec.yaml	2019/3/30 18:18	YAML 文件	3 KB
README.md	2019/3/18 22:10	MD 文件	1 KB

图 4-2　在项目根目录下创建多个类别资源文件夹

以上两种资源目录创建方式都可以使用，需要根据实际需求进行选择。这里要注意的是，对于已定义的资源目录，我们需要先在 pubspec.yaml 文件中进行路径配置，然后才可以在 Flutter 代码里使用，如图 4-3 所示。

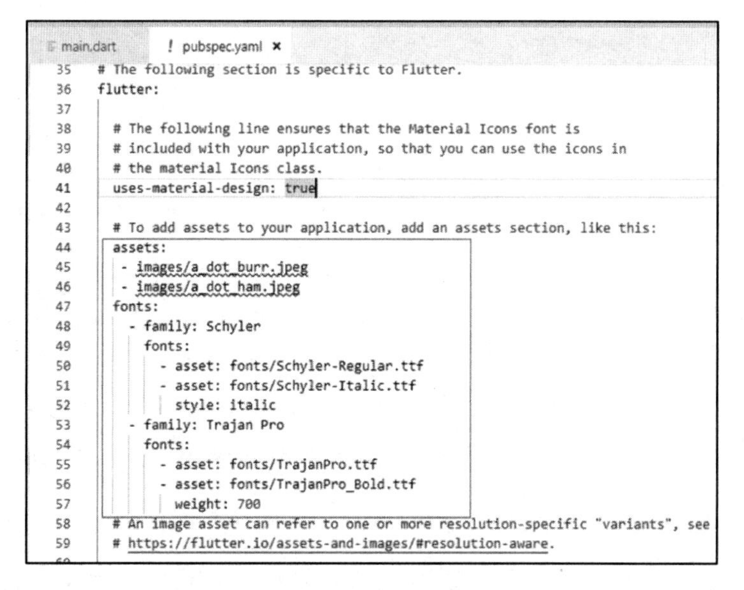

```
main.dart    ! pubspec.yaml ×
35  # The following section is specific to Flutter.
36  flutter:
37
38    # The following line ensures that the Material Icons font is
39    # included with your application, so that you can use the icons in
40    # the material Icons class.
41    uses-material-design: true
42
43    # To add assets to your application, add an assets section, like this:
44    assets:
45     - images/a_dot_burr.jpeg
46     - images/a_dot_ham.jpeg
47    fonts:
48      - family: Schyler
49        fonts:
50          - asset: fonts/Schyler-Regular.ttf
51          - asset: fonts/Schyler-Italic.ttf
52            style: italic
53      - family: Trajan Pro
54        fonts:
55          - asset: fonts/TrajanPro.ttf
56          - asset: fonts/TrajanPro_Bold.ttf
57            weight: 700
58    # An image asset can refer to one or more resolution-specific "variants", see
59    # https://flutter.io/assets-and-images/#resolution-aware.
```

图 4-3　路径配置

前面讲过，main.dart 文件位于 lib 目录下，是整个应用的入口文件，名称不可以修改，位置也不可以修改。在 lib 目录下创建其他类时，可以按照功能来创建，如创建 api、conf、ui、utils、widgets 等类，lib 目录结构如图 4-4 所示。

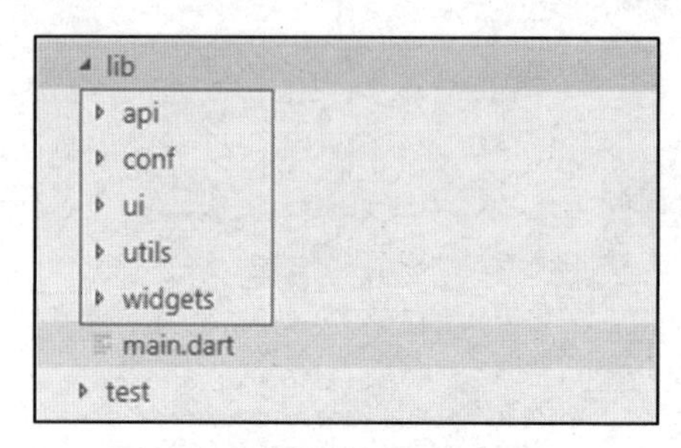

图 4-4　lib 目录结构

如果想引入第三方库，可以在 Dart Pub 仓库中搜索，然后在 pubspec.yaml 文件中为其配置路径，如图 4-5 所示。

```
dependencies:
  flutter:
    sdk: flutter
  shared_preferences: ^0.5.1+2
  # The following adds the Cupertino Icons font to your application.
  # Use with the CupertinoIcons class for iOS style icons.
  cupertino_icons: ^0.1.2
  #引入第三方库

dev_dependencies:
  flutter_test:
    sdk: flutter
```

图 4-5　为第三方库配置路径

4.2　命名规范

大部分编程语言都有自己的命名特点，不过整体来说大同小异。本节我们介绍 Flutter 的命名规范。

Flutter 的命名规范其实就是 Dart 语言的命名规范，具体来说有以下三种方式。

- UpperCamelCase：单词首字母大写的驼峰式命名方式，如 StudentName。
- lowerCamelCase：第一个单词首字母小写的驼峰式命名方式，如 studentName。

- lowercase_with_underscores：单词全部小写，中间用下画线“_”连接的命名方式，如 student_name。

这三种命名方式一般都在哪种情况下使用呢？下面我们分别介绍。一般情况下，类、注解、枚举类型、typedef、参数类型使用 UpperCamelCase 方式命名，示例如下。

```
// 类名
class HttpApi { ... }

// 注解名
@Foo()

// 枚举类型
enum Color {
  LightRed,
  LightBlue
}

// typedef
typedef Predicate<T> = bool Function(T value);

// 参数类型
@override
Widget build(BuildContext context) {...}
```

lowerCamelCase 命名方式一般用在变量、方法和参数、常量等的命名上，示例如下。

```
// 变量
HttpRequest httpRequest;

// 方法和参数
void align(bool clearItems) {
  // ...
}

// 常量
const defaultTimeout = 1000;
```

最后我们来看一下 lowercase_with_underscores 命名方式。lowercase_with_underscores 命名方式一般用在库（libraries）、包（packages）、源文件（source files）的命名上，具体示例如下。

```
// 库命名
library json_parser.string_scanner;
// 包命名
import 'package:event_bus/event_bus.dart';
// 源文件命名
import 'file_system.dart';
import 'item_menu.dart';
```

同时，在 Flutter 导入类库的时候，as 关键字后面的命名也要采用 lowercase_with_underscores 方式，示例如下。

```
import 'package:angular_components/angular_components' as angular_components;
```

Flutter 命名时还有一点需要注意：不要使用前缀字母。

```
// 推荐
defaultTimeout
// 不推荐
kDefaultTimeout
```

为了保持代码的整洁及层次，我们可以在某些地方使用空行。

如果某个方法、常量、变量、类不想被其他外部类访问或调用的话，可以在相应的名称前加上下画线前缀，示例如下。

```
// 该类不能被其他类访问、调用
class _MyMainPageState extends State<MyMainApp> {
  @override
  void initState() {
    super.initState();
  }
  ...
```

4.3　代码格式规范

很多开发工具都自带格式化工具和插件，如在 VSCode 中可以使用 Alt+Shift+F 组合键进行代码格式化。本节我们来介绍 Flutter 代码格式规范。

Flutter 官方建议使用 dartfmt 插件进行代码格式化，非常方便。如果遇到格式化工具无法处理的代码，建议重新组织代码，如缩短局部变量名称、更改层级等。官方建议，每行代码不超过 80 个字符。

对于流程控制相关语句，要用花括号{}将流程内容语句括起来，示例如下。

```
if (isWeekDay) {
  print('Bike to work!');
} else {
  print('Go dancing or read a book!');
}
```

如果一个控制语句中只有 if，没有 else，则可以不使用{}，示例如下。

```
if (arg == null) return defaultValue;
```

但是，如果 if 里的判断语句和 return 的返回语句都很长，可能产生换行，则依然建议使用花括号，示例如下。

```
if (overflowChars != other.overflowChars) {
  return overflowChars < other.overflowChars;
}
```

其他需要注意的是，如果遇到一些层级嵌套很深的情况，可以将某个层级的内容定义为另一个方法进行调用和引入。

```
class _MyMainPageState extends State<MyMainApp> {
  @override
  void initState() {
    super.initState();
  }

  @override
  Widget build(BuildContext context) {
    return MaterialApp(
      home: Scaffold(
        appBar: AppBar(
          title: Text('标题'),
        ),
        // 通过调用方法引入布局
        body: getBody(),
      ),
    );
  }
  // 布局单独用一个方法定义，避免层级嵌套太深
  Widget getBody() {
    return Center(child: Text("我是内容"));
```

```
  }
}
```

4.4 注释规范

Flutter 注释分为很多种。首先来看一下"//"形式的单行注释，这种注释不会在文档里出现或生成，只在代码中出现，示例如下。

```
// 单行注释，不会在文档里出现、生成
if (_chunks.isEmpty) return false;
```

再来看一下块注释（多行注释），我们可以使用/* ... */来注释多行代码内容，示例如下。但是并不推荐用这种方式对单行内容进行注释。

```
// 注释多行代码内容
/*
class _MyMainPageState extends State<MyMainApp> {
  @override
  void initState() {
    super.initState();
  }*/

greet(name) {
  /* 用多行注释（块注释）方式进行单行注释是不建议的*/
  print('Hi, $name!');
}
```

最后来看一下文档注释，文档注释使用"///"来表示，并且注释内容会在文档里出现和生成。一般我们可以使用文档注释对类成员、类型、方法、参数、变量、常量等进行注释。

```
/// 获取字符长度
int get length => ...
```

以上示例不建议使用单行注释，而要使用文档注释。因为对方法进行说明时，使用文档注释可以在引用方法的时看到该方法的提示，便于理解方法的使用规则。还有一种多行注释方式也是被支持的，只不过 Flutter 中不建议使用，示例如下。

```
/**
 * 多行注释，不建议使用，但也是被支持的
 */
```

对于文档注释，我们也可以在其中加入一些代码，示例如下。

```
/// 注释内有代码
/// {@tool sample}
/// ```dart
/// AppBar(
///   leading: Builder(
///     builder: (BuildContext context) {
///       return IconButton(
///        …
///       );
///     },
///   ),
/// )
/// ```
/// {@end-tool}
/// 注释内容
```

另外，我们也可以在文档注释中适当引入方括号（[]），以强调某个变量、参数、类或其他内容，示例如下。

```
/// 将字符串 [text] 追加到原始字符串后面
void add(String text) {
  ...
}
```

Flutter 也支持在文档注释里加入 MarkDown 文本，代码如下。

```
/// 这是正常的文字
///
/// 以下为 MarkDown 文本
/// * MarkDown 符号
/// 1. MarkDown 列表
/// 2. MarkDown 列表
```

但是我们应该避免在代码中过度使用 MarkDown 文本，因为这样可能会导致文档注释非常混乱，不利于阅读。写注释时候，建议精炼简短。适当的时候可以用空行来分隔注释内容。另外，不要把注释内容和上下文代码混合在一起，以免造成阅读障碍。

4.5　代码使用规范

Flutter 的代码使用规范比较多，在本节中，我们将按照与包导入相关、与字符串相关、与集合相关、与函数相关、与异常处理相关、与异步任务编程相关、与数据转换相关这样的逻辑顺序来一一介绍。

4.5.1　与包导入相关的规范

要想学习 Flutter 中与包导入相关的代码使用规范，我们可以结合示例来理解。假设 Flutter 项目中的包结构如下。

```
my_package
└ lib
   ├ src
   │  └ utils.dart
   └ api.dart
```

现在我们想在 api.dart 文件中导入 scr 下的 utils.dart 类，建议方式与不建议方式分别如下。

```
// 建议引入相对路径
import 'src/utils.dart';

// 不建议
import 'package:my_package/src/utils.dart';
```

进行导入操作时，不需要加入 package，这是因为，若后续 package 的名字发生变化，修改起来将非常麻烦。

4.5.2　与字符串相关的规范

在 Flutter 中，字符串连接时不推荐用“+”，只需用变量引用的方式将需要连接的字符串挨着书写即可，示例如下。

```
// 将变量或字符串与其他字符串直接连接
'Hello, $name! You are ${year - birth} years old.';

// 不推荐使用“+”连接
'Hello, ' + name + '! You are ' + (year - birth).toString() + ' years old.';
```

4.5.3　与集合相关的规范

Flutter 中的集合类型有 List、Map、Queue、Set 几种，对于集合的操作也有很多种，如创建、判断、转换、遍历等。我们首先来看一下创建空集合的方法。Flutter 中建设尽量使用简单的字面量创建空集合，而无须重复写出它的数据类型。

```
// 建议
var points = [];
var addresses = {};

// 不建议
var points = List();
var addresses = Map();
```

接下来看一下提供类型参数声明的集合创建方法。如果需要将集合指定为某种类型，即泛型，直接在集合字面量前写上类型对象并用尖括号包裹即可，无须重复写出数据类型。

```
// 建议
var points = <Point>[];
var addresses = <String, Address>{};

// 不建议
var points = List<Point>();
var addresses = Map<String, Address>();
```

在判断集合是否为空时，建议使用 isEmpty 和 isNotEmpty 方法，而不要使用.length 方法，示例如下。

```
// 建议
if (lunchBox.isEmpty) return 'so hungry...';
if (words.isNotEmpty) return words.join(' ');

// 不建议
if (lunchBox.length == 0) return 'so hungry...';
if (!words.isEmpty) return words.join(' ');
```

对于集合转换，我们可以使用链式高级方法来实现，即使用.where()、.map()等集合自带的高阶方法进行转换，示例如下。

```
var aquaticNames = animals
    .where((animal) => animal.isAquatic)
    .map((animal) => animal.name);
```

在进行集合的循环遍历时，建议使用 for 方法，不建议使用 forEach 写法，示例如下。

```
// 建议
for (var person in people)

// 不建议
people.forEach((person) {
```

```
  ...
});
```

在进行类型转换时，以下两种方式都可以实现，但是推荐第一种写法，不建议使用 List.from 方法，示例如下。

```
// 建议
var iterable = [1, 2, 3];
print(iterable.toList().runtimeType);

// 不建议
var iterable = [1, 2, 3];
print(List.from(iterable).runtimeType);
```

但是如果要改变集合类型，可以使用 List.from 方法，以下示例将 Number 类型集合转为 int 类型集合。

```
var numbers = [1, 2.3, 4]; // List<num>.
numbers.removeAt(1);
var ints = List<int>.from(numbers);
```

关于集合过滤，有如下几种写法，但建议采用第一种简单写法，示例如下。

```
// 建议
var objects = [1, "a", 2, "b", 3];
var ints = objects.whereType<int>();

// 不建议
var objects = [1, "a", 2, "b", 3];
var ints = objects.where((e) => e is int);

// 不建议
var objects = [1, "a", 2, "b", 3];
var ints = objects.where((e) => e is int).cast<int>();
```

4.5.4　与函数相关的规范

我们在命名函数时，需要注意一些地方：能够使用精简方式定义的，建议都使用精简方式去定义，而无须使用冗余的语句进行定义，否则会导致代码可读性降低，也有可能创建不必要的对象，示例如下。

```
// 建议
```

```
void main() {
  localFunction() {
    ...
  }
}

// 不建议
void main() {
  var localFunction = () {
    ...
  };
}
```

同样地，在使用函数时，建议使用形式精简的调用方法，提高代码的可读性，而无须编写额外的冗余代码，示例如下。

```
// 建议
names.forEach(print);

// 不建议
names.forEach((name) {
  print(name);
});
```

在将函数中的默认值和参数进行分隔时，建议采用等号形式赋值，不建议使用冒号形式进行值传递，以免降低代码可读性和规范性，示例如下。

```
// 建议
void insert(Object item, {int at = 0}) { ... }
// 不建议
void insert(Object item, {int at: 0}) { ... }
```

在判断函数传递的参数是否为 null 时，可以使用“??”来判断，示例如下。

```
void error([String message]) {
  // 判断 message 参数是否为空，非空就输出内容
  stderr.write(message ?? '\n');
}
```

在对函数进行处理时，还需要注意，不要将变量初始化为 null，而应该使用其默认值，或者为其赋一个初始值，这也是代码开发规范的重要细节。如果将变量赋值为 null，而真正使用变量时忘记初始化，则会出现应用空指针情况，示例如下。

```
// 建议使用默认值
int _nextId;
// 或赋初始值
int _nextId = 0;

// 不建议
int _nextId = null;
```

在对函数进行处理时，可以使用 final 来创建只读常量，也支持“=>”简写函数返回体方式，示例如下。

```
class Box {
  final contents = [];
}

double get area => (right - left) * (bottom - top);
// 使用=>省略了{...}和 return
```

在对函数传递的参数赋值时可以使用 this 关键字，但是不建议重复多次使用 this 关键字，这是没有必要的。例如在类中调用明确的方法时，代码如下。

```
// 建议
class Box {
  var value;

  void clear() {
    update(null);
  }
  void update(value) {
    this.value = value;
  }
}

// 不建议多次使用 this
class Box {
  var value;

  void clear() {
    this.update(null);
  }
  void update(value) {
```

```
    this.value = value;
  }
}
```

我们要尽量在声明中对常量进行初始化，以避免忘记，因为未进行初始化的常量没有任何意义，而且会导致空指针形成，示例如下。

```
class Folder {
  final String name;
  final List<Document> contents = [];

  Folder(this.name);
  // 进行常量初始化
  Folder.temp() : name = 'temporary';
}
```

我们可以使用 Dart 语言的特点来缩减构造方法参数的初始化写法，这样代码会更加简单，可读性更强，示例如下。

```
class Point {
  num x, y;
  Point(this.x, this.y);
}
```

在构造方法里，我们无须重复声明参数类型，直接赋值即可。重复声明没有任何必要，这不是规范的写法，而且会导致系统资源浪费和可读性降低，示例如下。

```
// 建议
class Point {
  int x, y;
  Point(this.x, this.y);
}
// 不建议，无须重复声明 int 类型
class Point {
  int x, y;
  Point(int this.x, int this.y);
}
```

对于空方法体的构造方法，直接在结尾处写“;”即可，无须将后面的空方法体写全，否则代码会很烦琐，不易于阅读，示例如下。

```
// 建议
class Point {
```

```
  int x, y;
  Point(this.x, this.y);
}
// 不建议
class Point {
  int x, y;
  Point(this.x, this.y) {}
}
```

在对函数进行传参和初始化填充数据时，无须重复定义 const 关键字，重复定义没有任何意义，并且会影响代码可读性，示例如下。

```
// 建议
const primaryColors = [
  Color("red", [255, 0, 0]),
  Color("green", [0, 255, 0]),
];
// 不建议
const primaryColors = const [
  const Color("red", const [255, 0, 0]),
  const Color("green", const [0, 255, 0]),];
```

4.5.5 与异常处理相关的规范

简单来说，在 Flutter 中，我们可以使用 rethrow 抛出代码异常，以供后续逻辑处理，示例如下。

```
try {
  somethingRisky();
} catch (e) {
  if (!canHandle(e)) rethrow;
  handle(e);
}
```

4.5.6 与异步任务编程相关的规范

我们可以使用 Future、async 和 await 来处理异步编程，其中 async 和 await 要成对出现，示例如下。

```
Future<int> countActivePlayers(String teamName) async {
```

```
  try {
    var team = await downloadTeam(teamName);
    if (team == null) return 0;

    var players = await team.roster;
    return players.where((player) => player.isActive).length;
  } catch (e) {
    log.error(e);
    return 0;
  }
}
```

如果有些方法中没有用到异步任务，则不要使用 async 关键字，示例如下。

```
// 使用 async 关键字是没有必要的
Future afterTwoThings(Future first, Future second) async {
  return Future.wait([first, second]);
}
```

4.5.7　与数据转换相关的规范

关于数据转换，我们可以使用 Future 里的高级用法进行简化。例如有些逻辑功能已经在 Future 里封装好了，我们只需要直接调用即可，无须重新实现。这样做可以让代码可读性更高，并且可以有效避免创建无关的对象或变量，示例如下。

```
// 建议，链式调用，代码精简，可读性高，避免创建无关对象
Future<bool> fileContainsBear(String path) {
  return File(path).readAsString().then((contents) {
    return contents.contains('bear');
  });
}
// 建议，链式调用，代码精简，可读性高，避免创建无关对象
Future<bool> fileContainsBear(String path) async {
  var contents = await File(path).readAsString();
  return contents.contains('bear');
}
// 不建议，分开编写冗余逻辑，代码可读性低，创建无关对象
Future<bool> fileContainsBear(String path) {
  var completer = Completer<bool>();
```

```
  File(path).readAsString().then((contents) {
    completer.complete(contents.contains('bear'));
  });

  return completer.future;
}
```

在进行不确定的函数数据类型封装时，可以适当地使用 T 泛型，这样函数对象就可以接收通用类型对象的传入，使函数对象更具通用性，示例如下。

```
Future<T> logValue<T>(FutureOr<T> value) async {
  if (value is Future<T>) {
    var result = await value;
    print(result);
    return result;
  } else {
    print(value);
    return value as T;
  }
}
```

掌握良好的代码开发规范非常重要，不但有利于提升自己的开发效率和编程水平，也能让其他人更好理解我们编写的代码。好的代码规范对于团队合作、项目可读性、迭代更新、安全性、可维护性都具有正向的催进作用。建议大家认真学习这部分内容，为后续的 Flutter 开发奠定基础。

第5章

Flutter常用组件（上）

前面已经讲解了大量Flutter基础知识，从本章开始，我们将介绍Flutter常用组件。我们知道，Flutter中几乎所有的对象都可以看成组件，这些组件不单单是UI控件，也具备一些逻辑操作功能，因此本章非常重要。

本章将主要讲解文本类组件、图片类组件、导航类组件。

5.1 文本类组件

顾名思义，文本类组件就是实现文件显示或输入等功能的组件，Flutter 中的文本类组件用途也如此。本节介绍的文本类组件主要有Text组件、Button组件、TextField组件。

5.1.1 Text组件

Text组件的作用是，在Flutter里显示文本信息，功能类似于Android的TextView，属于基础组件。Text组件的继承关系如下。

```
Text -> StatelessWidget -> Widget -> DiagnosticableTree -> Diagnosticable -> Object
```

可以看出，Text组件继承自StatelessWidget，属于无状态组件。接下来看一下Text组件的类结构，如图5-1所示。

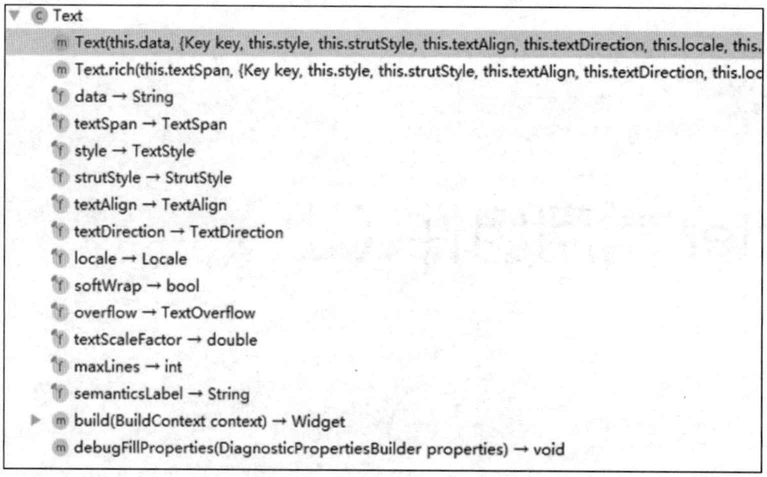

Text
Text(this.data, {Key key, this.style, this.strutStyle, this.textAlign, this.textDirection, this.locale, this.
Text.rich(this.textSpan, {Key key, this.style, this.strutStyle, this.textAlign, this.textDirection, this.loc
data → String
textSpan → TextSpan
style → TextStyle
strutStyle → StrutStyle
textAlign → TextAlign
textDirection → TextDirection
locale → Locale
softWrap → bool
overflow → TextOverflow
textScaleFactor → double
maxLines → int
semanticsLabel → String
build(BuildContext context) → Widget
debugFillProperties(DiagnosticPropertiesBuilder properties) → void

图 5-1　Text 组件的类结构

通过图 5-1 可以看出，Text 组件有两个同名构造方法，有多个属性参数（图 5-1 中前方有 f 图标标志的为属性参数，如 data、textSpan 等）。

下面我们来重点看一下 Text 组件的两个重要构造方法的属性参数含义。首先我们看一下 Text 组件第一个同名构造方法 Text 的属性参数含义，如下。

```
const Text(
    // 要显示的文字内容
    this.data,
   {
    // key类似于id
    Key key,
    // 文字显示样式和属性
    this.style,
    this.strutStyle,
    // 文字对齐方式
    this.textAlign,
    // 文字显示方向
    this.textDirection,
    // 语言环境
    this.locale,
    // 是否自动换行
    this.softWrap,
    // 文字溢出后的处理方式
```

```
    this.overflow,
    // 文字缩放
    this.textScaleFactor,
    // 最多显示行数
    this.maxLines,
    // 图像的语义描述，用于向 Android 上的 TalkBack 和 iOS 上的 VoiceOver 提供图像描述
    this.semanticsLabel,
  })
```

在以上代码中，data 属性是非空的，必须有参数，参数是要显示的 String 类型字符串。Text 组件里的 style 属性比较常用，传入的是 TextStyle 对象，TextStyle 可以配置的属性样式如下。

```
const TextStyle({
    // 判断是否继承父类组件属性
    this.inherit = true,
    // 文字颜色
    this.color,
    // 文字大小，默认为 14px
    this.fontSize,
    // 文字粗细
    this.fontWeight,
    // 字体，normal 或 italic
    this.fontStyle,
    // 字母间距，默认为 0，负数表示间距缩小，正数表示间距增大
    this.letterSpacing,
    // 单词间距，默认为 0，负数表示间距缩小，正数表示间距增大
    this.wordSpacing,
    // 文字基线
    this.textBaseline,
    // 行高
    this.height,
    // 设置区域
    this.locale,
    // 前景色
    this.foreground,
    // 背景色
    this.background,
    // 阴影
    this.shadows,
    // 下画线等装饰
```

```
    this.decoration,
    // 线条颜色
    this.decorationColor,
    // 线条样式，虚线、实线等
    this.decorationStyle,
    // 描述信息
    this.debugLabel,
    // 字体文件
    String fontFamily,
    List<String> fontFamilyFallback,
    String package,
  })
```

接下来再来看一下另一个构造方法 Text.rich 。这个方法的作用是在 Text 组件里加入一些 Span 标签，对某部分文字进行个性化设置，如加入@符号，加入超链接、表情等。Text.rich 等价于 RichText，两者的作用和用法基本一致，用哪个都可以。Text.rich 构造方法和 RichText 构造方法的属性参数如下。

```
// Text.rich()
const Text.rich(
    // 样式标签 TextSpan
    this.textSpan,
  {
    Key key,
    this.style,
    this.strutStyle,
    this.textAlign,
    this.textDirection,
    this.locale,
    this.softWrap,
    this.overflow,
    this.textScaleFactor,
    this.maxLines,
    this.semanticsLabel,
  })

// RichText()
const RichText({
    Key key,
    // 样式标签 TextSpan
```

```
    @required this.text,
    this.textAlign = TextAlign.start,
    this.textDirection,
    this.softWrap = true,
    this.overflow = TextOverflow.clip,
    this.textScaleFactor = 1.0,
    this.maxLines,
    this.locale,
    this.strutStyle,
  })
```

接下来我们看一下样式标签 TextSpan，其构造方法如下。

```
const TextSpan({
    // 样式片段
    this.style,
    // 要显示的文字
    this.text,
    // TextSpan 数组，可以包含多个 TextSpan
    this.children,
    // 用于手势识别，如点击跳转
    this.recognizer,
  })
```

关于 Text 组件的大部分功能和属性都介绍完了，接下来我们通过一个示例来演示 Text 组件的用法，代码如下。

```
// 以下为部分核心代码
body: Container(
      child: Column(
        children: <Widget>[
          Text('Text 最简单用法'),
          Text('Text Widget',
              textAlign: TextAlign.center,
              style: TextStyle(
                fontSize: 18,
                decoration: TextDecoration.none,
              )),
          Text('放大加粗文字',
              textDirection: TextDirection.rtl,
              textScaleFactor: 1.2,
              textAlign: TextAlign.center,
```

```
            style: TextStyle(
              fontWeight: FontWeight.bold,
              fontSize: 18,
              color: Colors.black,
              decoration: TextDecoration.none,
            )),
        Text(
            '可以缩放自动换行的文字，可以缩放自动换行的文字，可以缩放自动换行的文字，可以缩放自动
换行的文字…',
            textScaleFactor: 1.0,
            textAlign: TextAlign.center,
            softWrap: true,
            // 渐隐、省略号、裁剪
            maxLines: 2,
            overflow: TextOverflow.ellipsis,
            style: TextStyle(
              fontWeight: FontWeight.bold,
              fontSize: 18,
              color: Colors.black,
              decoration: TextDecoration.none,
            )),
        Text.rich(TextSpan(
          text: 'TextSpan',
          style: TextStyle(
            color: Colors.orange,
            fontSize: 30.0,
            decoration: TextDecoration.none,
          ),
          children: <TextSpan>[
            new TextSpan(
              text: '拼接1',
              style: new TextStyle(
                color: Colors.teal,
              ),
            ),
            new TextSpan(
              text: '拼接2',
              style: new TextStyle(
                color: Colors.teal,
              ),
```

```
          ),
          new TextSpan(
            text: '拼接 3 有点击事件',
            style: new TextStyle(
              color: Colors.yellow,
            ),
            recognizer: new TapGestureRecognizer()
              ..onTap = () {
                // 增加一个点击事件
                print(
                    '@@@@@@@@@@@@@@@@@@@@@@@@@@@@@@@@');
              },
          ),
        ],
      )),
      RichText(
        text: TextSpan(
          text: 'Hello ',
          style: DefaultTextStyle.of(context).style,
          children: <TextSpan>[
            TextSpan(
                text: 'bold',
                style: TextStyle(
                    fontWeight: FontWeight.bold,
                    decoration: TextDecoration.none)),
            TextSpan(
                text: ' world!',
                style: TextStyle(
                    fontWeight: FontWeight.bold,
                    decoration: TextDecoration.none)),
          ],
        ),
      ),
    ],
  ),
),
```

运行以上代码，可以看到 Text 组件效果如图 5-2 所示。

图 5-2　Text 组件效果

5.1.2　Button 组件

在 Flutter 中，Button 组件的作用是实现按钮功能，类似于 Android 的 Button、HTML 中的 button 标签，属于基础组件。

Flutter 中有很多封装好的 Button 组件：FlatButton（扁平化）、RaisedButton（有按下状态）、OutlineButton（有边框）、MaterialButton（Material 风格）、RawMaterialButton（没有应用 style 的 Material 风格按钮）、FloatingActionButton（悬浮按钮）、BackButton（返回按钮）、IconButton（Icon 图标）、CloseButton（关闭按钮）、ButtonBar（可以排列放置按钮元素的）等。

下面我们来看几个常用 Button 组件的效果，分别如图 5-3、图 5-4、图 5-5、图 5-6 所示。

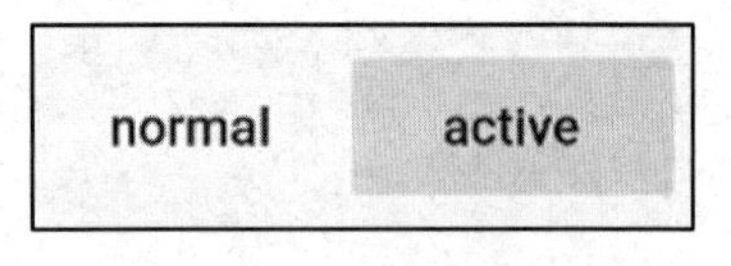

图 5-3　FlatButton 组件效果

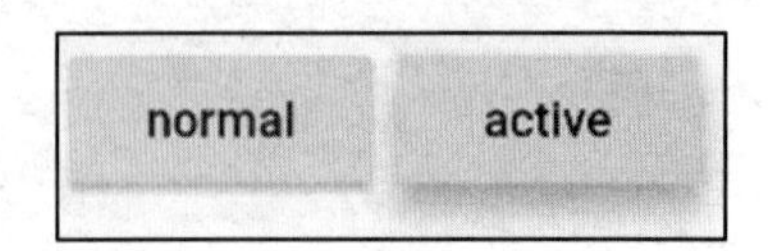

图 5-4　RaisedButton 组件效果

图 5-5　OutlineButton 组件效果

图 5-6　IconButton 组件效果

这里我们以比较常用的 FlatButton 组件为例，来分析一下它的构造方法，以及属性参数的含义，如下。

```
const FlatButton({
    Key key,
    // 点击事件
    @required VoidCallback onPressed,
    // 高亮显示，按下和抬起时都会调用
    ValueChanged<bool> onHighlightChanged,
    // 定义按钮的基色、最小尺寸、内部填充和默认形状
    ButtonTextTheme textTheme,
    // 按钮文字颜色
    Color textColor,
    // 按钮禁用时的文字颜色
    Color disabledTextColor,
    // 按钮背景颜色
    Color color,
    // 按钮禁用时的背景颜色
    Color disabledColor,
    // 按钮按下时的背景颜色
    Color highlightColor,
    // 按钮按下时，水波动画中水波的颜色（不要水波效果设置透明即可）
    Color splashColor,
    // 按钮主题，默认是浅色主题，也有深色
    Brightness colorBrightness,
    // 按钮填充间距
    EdgeInsetsGeometry padding,
    // 按钮外形
```

```
    ShapeBorder shape,
    Clip clipBehavior = Clip.none,
    MaterialTapTargetSize materialTapTargetSize,
    // 按钮内容，里面可以有子组件
    @required Widget child,
  })
```

接下来我们演示实现一个 Button 组件的过程，同时查看效果，代码如下。

```
body: CustomScrollView(
        slivers: <Widget>[
          SliverList(
            delegate: SliverChildListDelegate(<Widget>[
              Center(
                child: Column(
                  children: <Widget>[
                    // 用于返回的 Button
                    BackButton(
                      color: Colors.orange,
                    ),
                    // 用于关闭的 Button
                    CloseButton(),
                    ButtonBar(
                      children: <Widget>[
                        // 扁平化的 Button
                        FlatButton(
                          child: Text('FLAT BUTTON',
                              semanticsLabel: 'FLAT BUTTON 1'),
                          onPressed: () {
                            // 点击事件
                          },
                        ),
                      ],
                    ),
                    // 可以使用图标的 Button
                    FlatButton.icon(
                      disabledColor: Colors.teal,
                      label:
                          Text('FLAT BUTTON', semanticsLabel: 'FLAT BUTTON 2'),
                      icon: Icon(Icons.add_circle_outline, size: 18.0),
                      onPressed: () {},
```

```
),
ButtonBar(
  mainAxisSize: MainAxisSize.max,
  children: <Widget>[
    // 有边框轮廓的 Button
    OutlineButton(
      onPressed: () {},
      child: Text('data'),
    ),
  ],
),
ButtonBar(
  children: <Widget>[
    // 有图标、有边框轮廓的 Button
    OutlineButton.icon(
      label: Text('OUTLINE BUTTON',
          semanticsLabel: 'OUTLINE BUTTON 2'),
      icon: Icon(Icons.add, size: 18.0),
      onPressed: () {},
    ),
  ],
),
ButtonBar(
  children: <Widget>[
    // 按下时有水波动画的 Button
    RaisedButton(
      child: Text('RAISED BUTTON',
          semanticsLabel: 'RAISED BUTTON 1'),
      onPressed: () {
        // 点击事件
      },
    ),
  ],
),
ButtonBar(
  children: <Widget>[
    // 按下时有水波动画、有图标的 Button
    RaisedButton.icon(
      icon: const Icon(Icons.add, size: 18.0),
```

```
        label: const Text('RAISED BUTTON',
            semanticsLabel: 'RAISED BUTTON 2'),
        onPressed: () {
          // 点击事件
        },
      )
    ],
  ),
  ButtonBar(
    children: <Widget>[
      // Material 风格 Button
      MaterialButton(
        child: Text('MaterialButton1'),
        onPressed: () {
          // 点击事件
        },
      )
    ],
  ),
  ButtonBar(
    children: <Widget>[
      // 原始 Button
      RawMaterialButton(
        child: Text('RawMaterialButton1'),
        onPressed: () {
          // 点击事件
        },
      ),
    ],
  ),
  ButtonBar(
    children: <Widget>[
      // 悬浮的 Button
      FloatingActionButton(
        child: const Icon(Icons.add),
        heroTag: 'FloatingActionButton1',
        onPressed: () {
        },
        tooltip: 'floating action button1',
```

```
                            ),
                          ],
                        ),
                      ],
                    ),
                  ),
                ]),
              ),
            ],
          ),
```

执行以上代码，效果如图 5-7 所示。

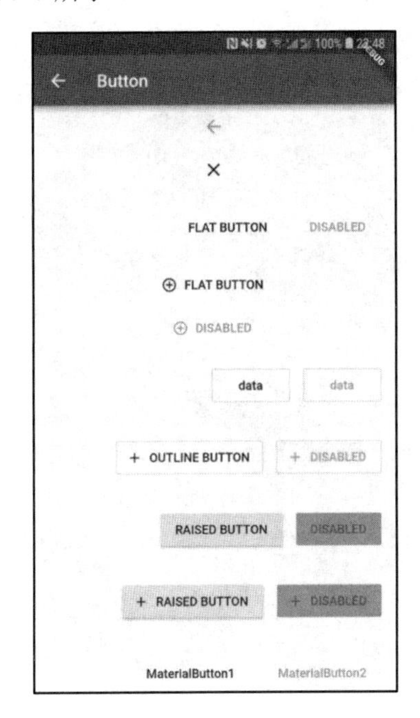

图 5-7　Button 组件效果

5.1.3　TextField 组件

Flutter 中的 TextField 组件相当于 Android 里的 EditText 或 HTML 的输入框。不过 TextField 组件可配置的选项更多一些，功能也更强大一些。

输入框一般都需要具有以下功能：提示信息、接收数据类型、监听事件等。Flutter 里的 TextField 组件同样具有这些功能。

Flutter 中有两个用于文本输入的 TextField 类型组件：一个是同名 TextField 组件，另一个是 TextFormField 组件。顾名思义，TextFormField 主要用于对 Form 表单进行操作。以下是 TextField 组件的构造方法及其属性参数的含义。

```
const TextField({
    Key key,
    // 输入框的控制器
    this.controller,
    // 焦点控制，控制是否获取或取消焦点
    this.focusNode,
    // 输入框的装饰类（重要）
    this.decoration = const InputDecoration(),
    // 输入框输入类型，键盘显示类型
    TextInputType keyboardType,
    // 控制键盘上的动作按钮图标
    this.textInputAction,
    // 控制键盘大小写显示
    this.textCapitalization = TextCapitalization.none,
    // 输入文本的样式
    this.style,
    // 输入文本的对齐方式
    this.textAlign = TextAlign.start,
    // 输入文本的方向
    this.textDirection,
    // 是否自动获取焦点
    this.autofocus = false,
    // 是否为密码（是否遮盖输入内容）
    this.obscureText = false,
    // 是否自动更正
    this.autocorrect = true,
    // 最大行数
    this.maxLines = 1,
    // 输入文本的最大长度
    this.maxLength,
    // 达到最大长度后，true 表示阻止输入，false 表示不阻止输入，但输入框会变红提示
    this.maxLengthEnforced = true,
    // 输入内容改变时的回调
    this.onChanged,
    // 输入完成，按回车键时调用
```

```
    this.onEditingComplete,
    // 输入完成，按回车键时回调，有回调参数，参数为输入内容
    this.onSubmitted,
    // 输入格式校验，如只能输入手机号
    this.inputFormatters,
    // 判断是否可用
    this.enabled,
    // 光标宽度
    this.cursorWidth = 2.0,
    // 光标圆角
    this.cursorRadius,
    // 光标颜色
    this.cursorColor,
    // 键盘样式，有暗黑主题、亮色主题
    this.keyboardAppearance,
    this.scrollPadding = const EdgeInsets.all(20.0),
    this.dragStartBehavior = DragStartBehavior.down,
    // 长按显示系统粘贴板内容
    this.enableInteractiveSelection,
    // 点击事件
    this.onTap,
    this.buildCounter,
  })
```

通过以上内容可知，TextField 组件中的构造方法有许多属性，分别代表不同的意义，下面我们着重介绍其中几个重要属性。

1. InputDecoration

InputDecoration 属性主要负责对输入框进行美化、装饰，以及对一些常用的样式进行配置。通过 InputDecoration 属性，我们可以个性化定制和配置 TextField 组件，使用起来非常方便，功能也非常强大。

2. TextInputType

TextInputType 属性主要用来控制输入内容的类型，如 text、multiline、number、phone、datetime、emailAddress、url，其中 number 还可以详细设置为 signed 或 decimal，这些输入类型和其他平台的输入类型基本一样。

3．controller

controller 属性一般用于获取输入框值、清空输入框、选择文本、监听等。

4．focusNode

focusNode 属性用于控制输入框焦点，例如我们按下回车键时，可以让某个输入框获取焦点，或监听焦点变化。

接下来我们就来看一下最基本的 TextField 组件的实现，代码如下。

```
TextField(
    decoration: InputDecoration(
        hintText: "输入用户名",
        icon: Icon(Icons.person)
    ),
)
```

执行以上代码，效果如图 5-8 所示。

图 5-8　TextField 组件效果

怎么样，实现一个 TextField 组件是不是很简单？接下来我们在此基础上编写一个简单的登录页面，代码如下。

```
@override
  Widget build(BuildContext context) {
    return Scaffold(
      appBar: AppBar(title: Text('TextField Widget'), primary: true),
      body: Column(
        children: <Widget>[
          Padding(
              padding: EdgeInsets.all(10),
              child: TextField(
                maxLines: 1,
                keyboardType: TextInputType.emailAddress,
```

```
            decoration: InputDecoration(
                border: OutlineInputBorder(
                    borderRadius: BorderRadius.all(Radius.circular(5.0)),
                    borderSide: BorderSide(color: Colors.grey)),
                labelText: '用户名',
                hintText: "输入用户名",
                icon: Icon(Icons.email)),
          )),
      Padding(
          padding: EdgeInsets.all(10),
          child: TextField(
            textInputAction: TextInputAction.done,
            maxLines: 1,
            keyboardType: TextInputType.text,
            obscureText: true,
            decoration: InputDecoration(
                border: OutlineInputBorder(
                    borderRadius: BorderRadius.all(Radius.circular(5.0)),
                    borderSide: BorderSide(color: Colors.grey)),
                labelText: '密码',
                hintText: "输入密码",
                icon: Icon(Icons.lock)),
          )),
    ],
  ),
 );
}
```

执行以上代码，效果如图 5-9 所示。

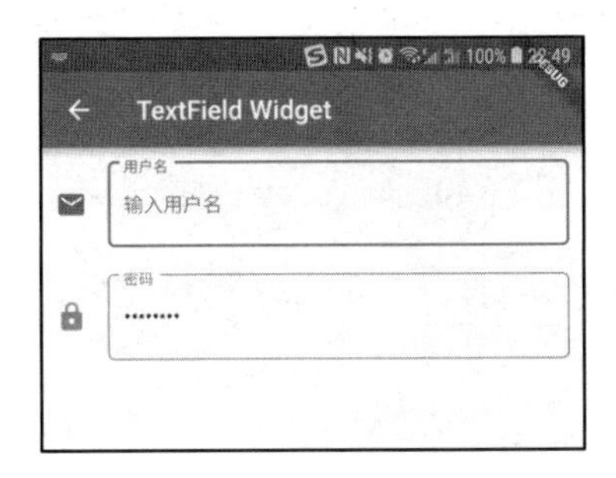

图 5-9　简单的登录页面

关于 TextField 组件的用法和相关属性就介绍到这里，大家可以尝试修改属性实现各种各样的组件效果。以上就是 Flutter 文本类组件的基本内容，内容覆盖比较全，基本上所有的属性都讲解到了，大家可以多多练习，举一反三。

5.2　图片类组件

顾名思义，图片类组件就是负责图片加载、显示、处理的组件。Flutter 中的图片类组件主要有 Image 组件和 Icon 组件两种，接下来我们一一介绍。

5.2.1　Image 组件

Image 组件在 Flutter 里是用来显示图片的，功能类似于 Android 的 ImageView、HTML 的 image 标签等，属于基础组件。Image 组件的继承关系如下。

```
Image -> StatefulWidget -> Widget -> DiagnosticableTree -> Diagnosticable -> Object。
```

Image 组件继承自 StatefulWidget，属于有状态组件，它的类结构如图 5-10 所示。

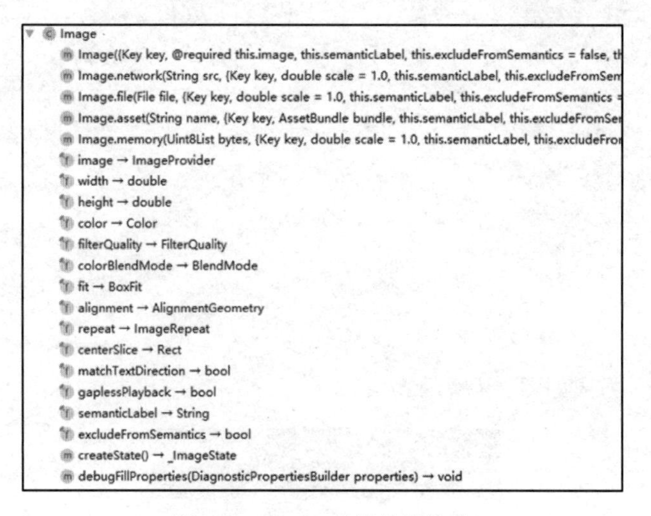

图 5-10　Image 组件的类结构

通过图 5-10 可以看出，Image 组件有 5 个构造方法，拥有多个属性参数，这 5 个构造方法可用于加载图片，具体如下。

- Image：通过 ImageProvider 来加载图片。
- Image.network：加载网络图片。

- Image.file：加载本地图片。
- Image.asset：加载项目内的资源图片。
- Image.memory：加载 Uint8List 资源图片或从内存中获取图片。

这几种构造方法其实都是通过 ImageProvider 从不同源来加载图片的，对应的封装类有 NetworkImage、FileImage、AssetImage、ExactAssetImage、MemoryImage，我们也可以使用这几个封装类来加载图片。

下面我们重点看一下 Image 组件的 5 个重要构造方法及其属性参数含义，具体如下。

```
// 通过 ImageProvider 来加载图片
const Image({
    Key key,
    // ImageProvider，图片显示源
    @required this.image,
    this.semanticLabel,
    this.excludeFromSemantics = false,
    // 显示宽度
    this.width,
    // 显示高度
    this.height,
    // 图片的混合色值
    this.color,
    // 混合模式
    this.colorBlendMode,
    // 缩放显示模式
    this.fit,
    // 对齐方式
    this.alignment = Alignment.center,
    // 重复方式
    this.repeat = ImageRepeat.noRepeat,
    // 当图片需要被拉伸显示时，通过 centerSlice 定义的矩形区域会被拉伸
    this.centerSlice,
    // 文字显示方向
    this.matchTextDirection = false,
    // 图片发生变化后，确认加载过程中原图片保留还是留白
    this.gaplessPlayback = false,
    // 图片显示质量
```

```
    this.filterQuality = FilterQuality.low,
  })

// 加载网络图片
Image.network(
    // 路径
    String src,
   {
    Key key,
    // 缩放
    double scale = 1.0,
    // 参考前面内容，不做重复说明
    this.semanticLabel,
    this.excludeFromSemantics = false,
    this.width,
    this.height,
    this.color,
    this.colorBlendMode,
    this.fit,
    this.alignment = Alignment.center,
    this.repeat = ImageRepeat.noRepeat,
    this.centerSlice,
    this.matchTextDirection = false,
    this.gaplessPlayback = false,
    this.filterQuality = FilterQuality.low,
    Map<String, String> headers,
  })

// 加载本地图片
Image.file(
    // File 对象
    File file,
  {
    Key key,
    // 不做重复说明
    double scale = 1.0,
    this.semanticLabel,
    this.excludeFromSemantics = false,
    this.width,
```

```
    this.height,
    this.color,
    this.colorBlendMode,
    this.fit,
    this.alignment = Alignment.center,
    this.repeat = ImageRepeat.noRepeat,
    this.centerSlice,
    this.matchTextDirection = false,
    this.gaplessPlayback = false,
    this.filterQuality = FilterQuality.low,
  })

// 加载项目内的资源图片
Image.asset(
    // 文件名称，包含路径
    String name,
  {
    Key key,
    // 用于访问资源对象
    AssetBundle bundle,
    // 不做重复说明
    this.semanticLabel,
    this.excludeFromSemantics = false,
    double scale,
    this.width,
    this.height,
    this.color,
    this.colorBlendMode,
    this.fit,
    this.alignment = Alignment.center,
    this.repeat = ImageRepeat.noRepeat,
    this.centerSlice,
    this.matchTextDirection = false,
    this.gaplessPlayback = false,
    String package,
    this.filterQuality = FilterQuality.low,
  })

// 加载 Uint8List 资源图片或从内存中获取图片
```

```
Image.memory(
    // Uint8List 资源图片
    Uint8List bytes,
  {
    Key key,
    // 不做重复说明
    double scale = 1.0,
    this.semanticLabel,
    this.excludeFromSemantics = false,
    this.width,
    this.height,
    this.color,
    this.colorBlendMode,
    this.fit,
    this.alignment = Alignment.center,
    this.repeat = ImageRepeat.noRepeat,
    this.centerSlice,
    this.matchTextDirection = false,
    this.gaplessPlayback = false,
    this.filterQuality = FilterQuality.low,
  })
```

以上代码给出了 Image 组件中 5 个重要构造方法的用法，下面我们来介绍混合模式和 fit 缩放显示模式。Flutter 中一共有 29 种混合模式，其中主要的混合模式效果如图 5-11 所示。

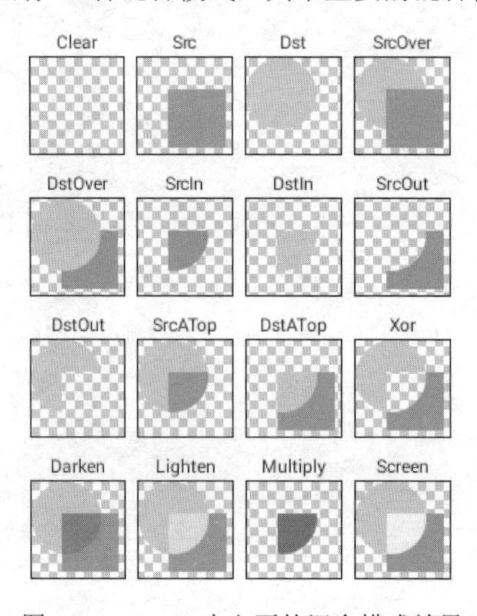

图 5-11　Flutter 中主要的混合模式效果

fit缩放显示模式主要有7种，分别是fill、contain、cover、fitWidth、fitHeight、none、scaleDown。下面我们分别展示各个模式的效果。

在 fill 模式下，图片会被拉伸，充满整个显示区域，效果如图 5-12 所示。

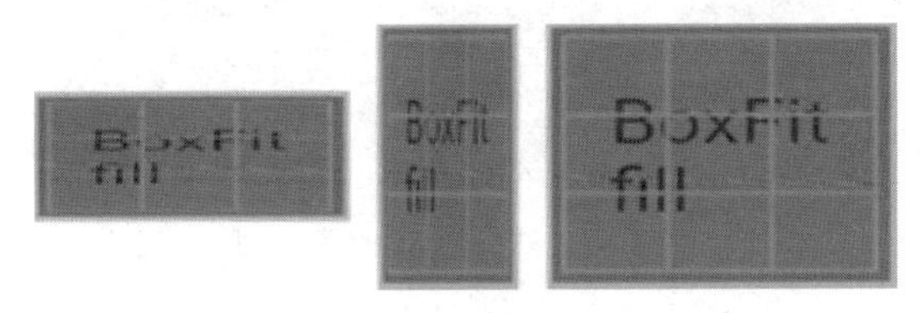

图 5-12 fill 模式效果

在 contain 模式下，图片会被全图显示且维持原比例，不会充满显示区域，效果如图 5-13 所示。

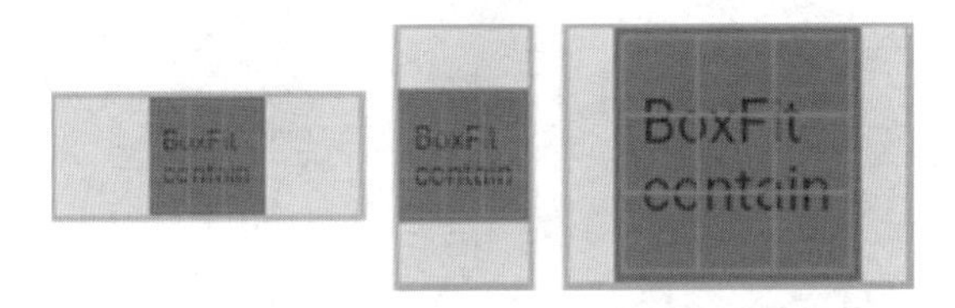

图 5-13 contain 模式效果

在 cover 模式下，图片可能会被拉伸，可能会被裁剪，也可能充满显示区域，效果如图 5-14 所示。

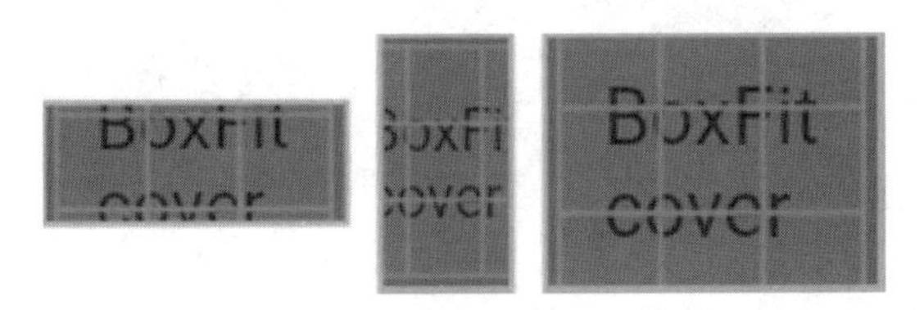

图 5-14 cover 模式效果

在 fitWidth 模式下，图片可能被拉伸，可能被裁剪，但是在宽度上是充满显示区域的，效果如图 5-15 所示。

图 5-15 fitWidth 模式效果

在 fitHeight 模式下，图片可能被拉伸，可能被裁剪，但是在高度上是充满显示区域的，效果如图 5-16 所示。

图 5-16　fitHeight 模式效果

在 none 模式下，图片不会被缩放，若超出显示区域就直接裁剪多余部分，其余部分保留，若不足显示区域则不处理，效果如图 5-17 所示。

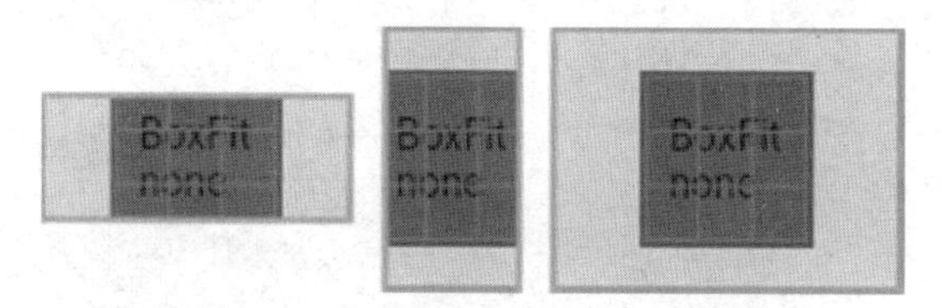

图 5-17　none 模式效果

在 scaleDown 模式下，图片会被全图显示且显示原比例，但不允许超过原图片大小，即图片会根据显示区域的大小进行缩放，可小不可大，效果如图 5-18 所示。

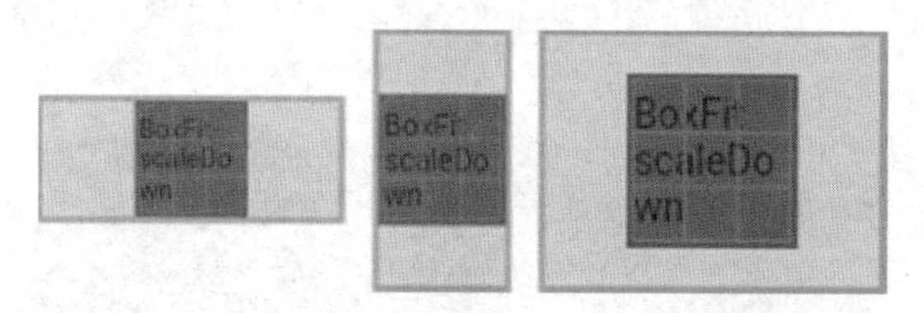

图 5-18　scaleDown 模式效果

以上介绍了 Image 组件的大部分功能和属性，接下来通过一个示例来演示 Image 组件的具体用法。

```
body: CustomScrollView(
        slivers: <Widget>[
          SliverPadding(
            padding: const EdgeInsets.all(20.0),
            sliver: SliverList(
              delegate: SliverChildListDelegate(
                <Widget>[
                  // 从项目目录里读取图片，需要在 pubspec.yaml 中注册路径
```

```
Image.asset("assets/assets_image.png"),
Image(
  image: AssetImage("assets/assets_image.png"),
  width: 200,
  height: 130,
),
// 从文件中读取图片
Image.file(
  File('/sdcard/img.png'),
  width: 200,
  height: 80,
),
Image(
  image: FileImage(File('/sdcard/img.png')),
),
/// 读取、加载原始图片
// RawImage(
//   image: imageInfo?.image,
// ),

/// 从内存中读取 byte 数组图片
/// Image.memory(bytes)
/// Image(
///   image: MemoryImage(bytes),
/// ),

// 读取网络图片
Image.network(imageUrl),
Image(
  image: NetworkImage(imageUrl),
),

/// 未成功加载图片时加入占位图
FadeInImage(
  placeholder: AssetImage("assets/assets_image.png"),
  image: FileImage(File('/sdcard/img.png')),
),
FadeInImage.assetNetwork(
  placeholder: "assets/assets_image.png",
```

```
    image: imageUrl,
  ),

  /// FadeInImage.memoryNetwork(
  ///   placeholder: byte,
  ///   image: imageUrL,
  /// ),

  /// 加载圆角图片
  CircleAvatar(
    backgroundColor: Colors.brown.shade800,
    child: Text("圆角头像"),
    backgroundImage: AssetImage("assets/assets_image.png"),
    radius: 50.0,
  ),

  ImageIcon(NetworkImage(imageUrl)),

  ClipRRect(
    child: Image.network(
      imageUrl,
      scale: 8.5,
      fit: BoxFit.cover,
    ),
    borderRadius: BorderRadius.only(
      topLeft: Radius.circular(20),
      topRight: Radius.circular(20),
    ),
  ),

  Container(
    width: 120,
    height: 60,
    decoration: BoxDecoration(
      shape: BoxShape.rectangle,
      borderRadius: BorderRadius.circular(10.0),
      image: DecorationImage(
          image: NetworkImage(imageUrl), fit: BoxFit.cover),
    ),
```

```
              ),

              ClipOval(
                child: Image.network(
                  imageUrl,
                  scale: 8.5,
                ),
              )
            ],
          ),
        ),
      ),
    ],
  ),
```

执行以上代码，效果如图 5-19 所示。通过以上程序，我们就生成了一个基本的 Image 组件。

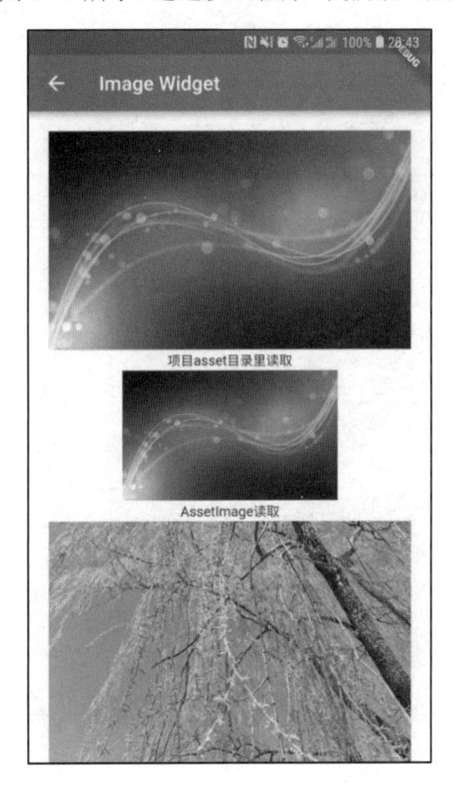

图 5-19　Image 组件效果

5.2.2 Icon 组件

Icon 组件的主要功能是显示与图标相关的内容。Flutter 中的 Icon 组件不但支持将常用的图片作为图标显示，也可以显示矢量图标、iconfont 里的图标，以及 Flutter 自带的 Material 风格图标，这时需要在 pubspec.yaml 中进行配置，具体如下。

```
// 配置以下内容，使用 Material 风格内置的 Icon 组件
flutter:

  uses-material-design: true
```

Material 风格图标通过“Icons.”来调用，如 Icons.add 表示可以使用加号图标。Flutter 的 Icon 组件中也封装了 ImageIcon 组件、IconButton 组件，可供开发者使用。所有 Material 风格图标都可以在官方网站查看，如图 5-20 所示。

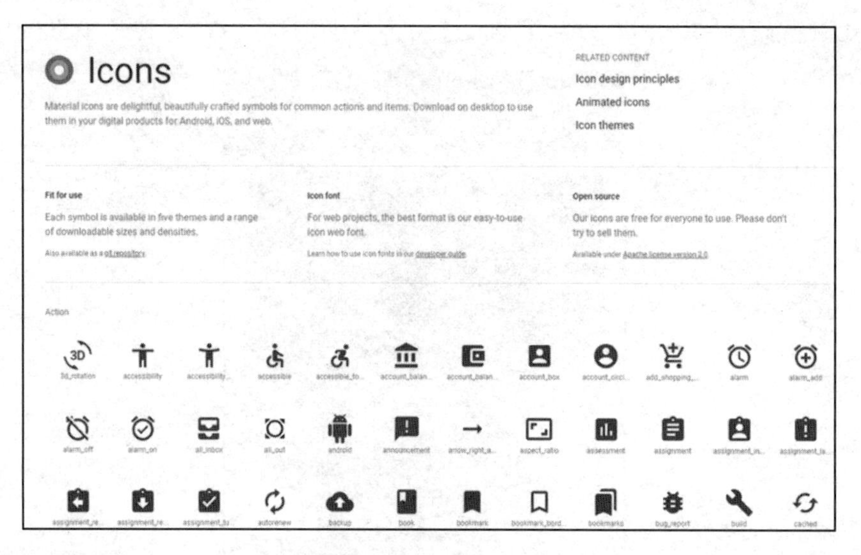

图 5-20 Material 风格图标

使用 Icon 组件可以带来许多好处和便利，具体来说有以下几点。

- 体积小：可以减小应用安装包体积。
- 矢量图标：iconfont 和自带的 Material 风格图标都是矢量图标，即使放大也不会影像图标清晰度。
- 可以动态改变参数：由于是 PNG 格式有透明度图标，因此可以改变颜色、大小、对齐方式等。

- 可以像表情一样通过将 TextSpan 组件和文本混用来展示。

Icon 组件也属于有状态组件，其构造方法如下。

```
const Icon(this.icon, {
  Key key,
  // 图标大小，默认为 24px
  this.size,
  // 图标颜色
  this.color,
  this.semanticLabel,
  // 图标方向，前提是将 IconData.matchTextDirection 字段设置为 true
  this.textDirection,
 })
```

接下来我们来看一个实现 Icon 组件的示例，代码如下。

```
class IconSamplesState extends State<IconSamples>
   with TickerProviderStateMixin {
 AnimationController _controller;

 @override
 void initState() {
   super.initState();

   /// 动画控制类，产生 0~1 的小数
   _controller = AnimationController(
       lowerBound: 0,
       upperBound: 1,
       duration: const Duration(seconds: 3),
       vsync: this);
       _controller.forward();
 }

 @override
 Widget build(BuildContext context) {
   return Scaffold(
     appBar: AppBar(
       title: Text('Icon Widget'),
       primary: true,
       leading: IconButton(
```

```
      icon: const Icon(Icons.menu),
      onPressed: () {},
    ),
  ),
  body: Column(
    children: <Widget>[
      IconButton(
        icon: Icon(Icons.directions_bike),
        // 处于按下状态时，按钮的主要颜色
        splashColor: Colors.teal,
        // 处于按下状态时，按钮的辅助颜色
        highlightColor: Colors.pink,
        // 按钮不可按下时的颜色
        disabledColor: Colors.grey,
        // PNG 图标主体颜色
        color: Colors.orange,
        onPressed: () {},
      ),
      // 图片为 PNG 格式，有透明度
      ImageIcon(
        AssetImage('assets/check-circle.png'),
        color: Colors.teal,
        size: 30,
      ),
      Icon(
        Icons.card_giftcard,
        size: 26,
      ),
      // 使用 unicode
      Text(
        "\uE000",
        style: TextStyle(
            fontFamily: "MaterialIcons",
            fontSize: 24.0,
            color: Colors.green),
      ),
      Icon(IconData(0xe614,
          // 也可以使用自定义字体
          fontFamily: "MaterialIcons",
```

```
            matchTextDirection: true)),
          AnimatedIcon(
            icon: AnimatedIcons.menu_arrow,
            progress: _controller,
            semanticLabel: 'Show menu',
          )
        ],
      ),
    );
  }
}
```

执行以上代码，效果如图 5-21 所示。

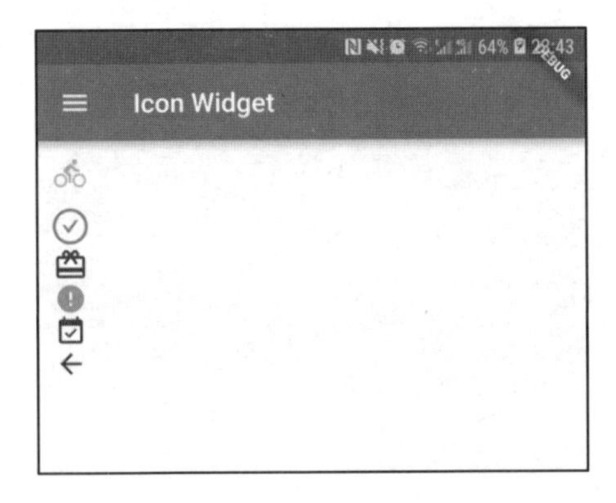

图 5-21　Icon 组件效果

5.3　导航类组件

导航类组件主要负责页面跳转、底部或顶部的操作菜单栏布局等，是很常用、很重要的基础组件。Flutter 中的导航类组件主要有以下几个：AppBar 组件、TabBar 组件、NavigationBar 组件、CupertinoTabBar 和 PageView 相关组件。

5.3.1　AppBar 组件

AppBar 组件主要用于实现页面的顶部标题栏，在里面定制各种按钮、菜单功能，一般配合 Scaffold 布局组件使用。AppBar 组件继承自 StatefulWidget，属于有状态组件，其构造方法如下。

```
AppBar({
  Key key,
```

```
  // 标题栏左侧图标，默认在首页显示 logo，在其他页面显示返回按钮，可以自定义
  this.leading,
  // 指明是否提供控件占位
  this.automaticallyImplyLeading = true,
  // 标题内容
  this.title,
  // 标题栏右侧菜单，可以用 IconButton 显示，也可以用 PopupMenuButton 显示为三个点
  this.actions,
  // AppBar 下方的控件，高度和 AppBar 一样，通常在 SliverAppBar 中使用
  this.flexibleSpace,
  // 标题栏下方的空间，一般放 TabBar
  this.bottom,
  // 标题栏阴影面积
  this.elevation,
  // 标题栏背景色
  this.backgroundColor,
  // 亮度
  this.brightness,
  // AppBar 上图标的颜色、透明度、尺寸信息
  this.iconTheme,
  // AppBar 上的文字样式
  this.textTheme,
  // 是否进入状态栏
  this.primary = true,
  // 标题是否居中
  this.centerTitle,
  // 标题间距，如果希望 title 占用所有可用空间，请将此值设置为 0.0
  this.titleSpacing = NavigationToolbar.kMiddleSpacing,
  // 透明度
  this.toolbarOpacity = 1.0,
  this.bottomOpacity = 1.0,
})
```

了解了 AppBar 组件的构造方法，我们来实现一个简单的 AppBar 组件，代码如下。

```
AppBar(
  title: Text('My Fancy Dress'),
  actions: <Widget>[
    IconButton(
      icon: Icon(Icons.playlist_play),
      tooltip: 'Air it',
```

```
      onPressed: _airDress,
    ),
    IconButton(
      icon: Icon(Icons.playlist_add),
      tooltip: 'Restitch it',
      onPressed: _restitchDress,
    ),
    IconButton(
      icon: Icon(Icons.playlist_add_check),
      tooltip: 'Repair it',
      onPressed: _repairDress,
    ),
  ],
)
```

执行以上代码，效果如图 5-22 所示。

图 5-22　简单的 AppBar 组件效果

掌握了方法，我们可以实现功能更丰富的 AppBar 组件，大家不妨试试！

除了 AppBar 组件，Flutter 还提供了一个 BottomAppBar 组件，可用于实现底部导航效果。这个组件的自定义能力非常强大，一般搭配 FloatingActionButton 使用。BottomAppBar 继承自 StatefulWidget，也是有状态组件，其构造方法如下。

```
const BottomAppBar({
    Key key,
    // 颜色
    this.color,
    this.elevation,
    // 设置底栏的形状
```

```
    this.shape,
    this.clipBehavior = Clip.none,
    this.notchMargin = 4.0,
    // 放置各种类型的组件，自定义性更强
    this.child,
  })
```

了解了构造方法，接下来我们通过代码来实现 BottomAppBar 组件示例。

```
class BottomAppBarState extends State<BottomAppBarSamples> {

  @override
  void initState() {
    super.initState();
  }

  @override
  Widget build(BuildContext context) {
    return Scaffold(
      appBar: AppBar(
        title: Text('BottomAppBar Demo'),
      ),
      // body 主体内容
      body: Text("Index 0: Home"),
      // 底部导航栏用 BottomAppBar 组件实现
      bottomNavigationBar: BottomAppBar(
        // 切口的距离
        notchMargin: 6,
        // 底部留出空缺
        shape: CircularNotchedRectangle(),
        child: Row(
          children: <Widget>[
            IconButton(
              icon: Icon(Icons.home),
              onPressed: null,
            ),
            IconButton(
              icon: Icon(Icons.business),
              onPressed: null,
            ),
            IconButton(
```

```
            icon: Icon(Icons.school),
            onPressed: null,
          ),
        ],
      ),
    ),
    floatingActionButton: FloatingActionButton(
        child: Icon(
          Icons.add,
          color: Colors.white,
        ),
        onPressed: null),
    floatingActionButtonLocation: FloatingActionButtonLocation.endDocked,
  );
 }
}
```

执行以上代码，效果如图 5-23 所示。

图 5-23　BottomAppBar 组件效果

5.3.2 TabBar 组件

TabBar 组件和 TabBarView 组件一般搭配使用，TabBar 组件可用来实现 Tab 导航部分，TabBarView 组件则用来实现 body 内容区域部分。TabBar 组件继承自 StatefulWidget，是一个有状态组件。TabBarView 组件也同样继承自 StatefulWidget，属于有状态组件。

首先我们来介绍 TabBar 组件的构造方法，如下。

```
const TabBar({
    Key key,
    // Tab 页组件集合
    @required this.tabs,
    // TabController 对象，控制 Tab 页
    this.controller,
    // 是否可以滚动
    this.isScrollable = false,
    // 指示器颜色
    this.indicatorColor,
    // 指示器高度
    this.indicatorWeight = 2.0,
    // 底部指示器的 Padding 间距
    this.indicatorPadding = EdgeInsets.zero,
    // 指示器装饰器
    this.indicator,
    // 指示器大小计算方式
    this.indicatorSize,
    // 选中 Tab 文字颜色
    this.labelColor,
    // 选中 Tab 文字样式
    this.labelStyle,
    // 每个 label 的 padding 值
    this.labelPadding,
    // 未选中 label 的颜色
    this.unselectedLabelColor,
    // 未选中 label 的样式
    this.unselectedLabelStyle,
    this.dragStartBehavior = DragStartBehavior.down,
    // 点击事件
    this.onTap,
```

```
  })
```

接下来我们再来看一下 TabBarVeiw 组件的构造方法，如下。

```
const TabBarView({
    Key key,
    // Tab 内容页列表,和 TabBar 组件的 Tab 数量一样
    @required this.children,
    // TabController 对象，控制 Tab 页
    this.controller,
    this.physics,
    this.dragStartBehavior = DragStartBehavior.down,
  })
```

了解了 TabBar 组件和 TabBarView 组件的构造方法，我们结合两者，实现一个具体例子，代码如下。

```
class TabBarSamplesState extends State<TabBarSamples>
    with SingleTickerProviderStateMixin {
  TabController _tabController;

  @override
  void initState() {
    super.initState();
    // initialIndex 指明初始时选中第几个页面，length 为数量
    _tabController = TabController(initialIndex: 0, length: 5, vsync: this);
    // 监听
    _tabController.addListener(() {
      switch (_tabController.index) {
        case 0:

          break;
        case 1:

          break;
      }
    });
  }

  @override
  Widget build(BuildContext context) {
    return Scaffold(
```

```
    appBar: AppBar(
      title: Text('TabBar Demo'),
      primary: true,
      // 设置 TabBar 组件
      bottom: TabBar(
        controller: _tabController,
        tabs: <Widget>[
          Tab(
            text: "Tab1",
          ),
          Tab(
            text: "Tab2",
          ),
          Tab(
            text: "Tab3",
          ),
          Tab(
            text: "Tab4",
          ),
          Tab(
            text: "Tab5",
          ),
        ],
      ),
    ),
    // body 内容用 TabBarView 组件实现
    body: TabBarView(
      controller: _tabController,
      children: <Widget>[
        Center(
          child: Text("TabBarView data1"),
        ),
        Center(
          child: Text("TabBarView data2"),
        ),
        Center(
          child: Text("TabBarView data3"),
        ),
        Center(
```

```
            child: Text("TabBarView data4"),
          ),
          Center(
            child: Text("TabBarView data5"),
          ),
        ],
      ),
    );
  }

  @override
  void dispose() {
    super.dispose();
    _tabController.dispose();
  }
}
```

执行以上代码，效果如图 5-24 所示。

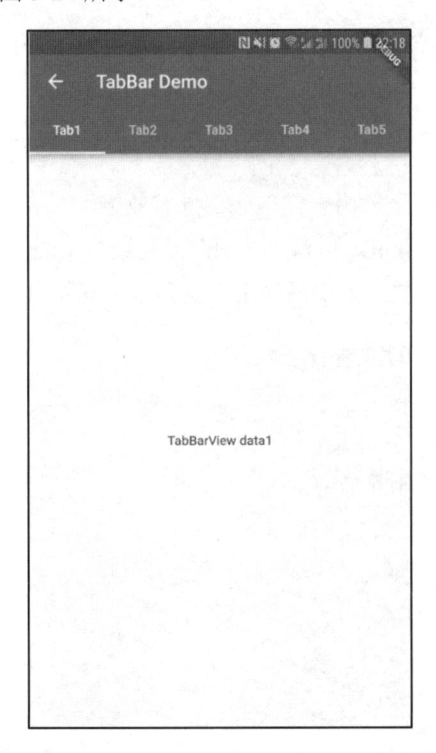

图 5-24　TabBar 组件结合 TabBarView 组件效果

如果想将 TarBar 组件放在底部，可以采用如下方式进行设置。

```
// 最外层是 Scaffold 布局组件
bottomNavigationBar: Material(
        color: Colors.blue,
        child: TabBar(
          controller: _controller,
          tabs: <Tab>[
            Tab(text: "Home", icon: Icon(Icons.home)),
            Tab(text: "Apps", icon: Icon(Icons.list)),
            Tab(text: "Center", icon: Icon(Icons.message)),
          ],
          indicatorWeight: 0.1,
        ),
      ),
```

5.3.3　NavigationBar 组件

NavigationBar 组件用于实现导航菜单按钮功能。例如很多应用界面的顶部或底部会有一个切换页面的导航菜单栏，这些效果就可以用 NavigationBar 组件来实现。Flutter 的 NavigationBar 组件里有一些导航组件，这里我们选择最常用的 BottomNavigationBar 组件来进行演示。

BottomNavigationBar 组件一般用来实现底部导航效果，继承自 StatefulWidget，属于有状态组件，一般搭配 BottomNavigationBarItem 组件使用。通过 BottomNavigationBar 组件实现的导航有一个特点：选项中会有一个放大的动画效果，这和通过其他组件实现的导航效果有所差别。

BottomNavigationBar 组件的构造方法如下。

```
BottomNavigationBar({
   Key key,
   // BottomNavigationBarItem 集合
   @required this.items,
   this.onTap,
   // 当前选中位置
   this.currentIndex = 0,
   // 设置显示模式
   BottomNavigationBarType type,
   // 主题色
   this.fixedColor,
```

```
  // 图标尺寸
  this.iconSize = 24.0,
})
```

接下来我们通过代码来实现一个 BottomNavigationBar 组件，具体如下。

```
... ...
class NavigationBarState extends State<NavigationBarSamples> {
  // 默认选中第一项
  int _selectedIndex = 0;

  final _widgetOptions = [
    Text('Index 0: Home'),
    Text('Index 1: Business'),
    Text('Index 2: School'),
  ];

  @override
  Widget build(BuildContext context) {
    return Scaffold(
      appBar: AppBar(
        title: Text('BottomNavigationBar Demo'),
      ),
      // 主体内容
      body: Center(
        child: _widgetOptions.elementAt(_selectedIndex),
      ),
      // 底部导航通过 BottomNavigationBar 组件实现
      bottomNavigationBar: BottomNavigationBar(
        items: <BottomNavigationBarItem>[
          // 单个 BottomNavigationBarItem 组件
          BottomNavigationBarItem(icon: Icon(Icons.home), title: Text('Home')),
          BottomNavigationBarItem(
              icon: Icon(Icons.business), title: Text('Business')),
          BottomNavigationBarItem(
              icon: Icon(Icons.school), title: Text('School')),
        ],
        // 选中位置
        currentIndex: _selectedIndex,
        // 主题色
```

```
      fixedColor: Colors.deepPurple,
      // 点击
      onTap: _onItemTapped,
    ),
  );
}

void _onItemTapped(int index) {
  setState(() {
    _selectedIndex = index;
  });
}
}
```

执行以上代码，效果如图 5-25 所示。

图 5-25　BottomNavigationBar 组件效果

前面提到，BottomNavigationBar 通常搭配 BottomNavigationBarItem 一起使用，这里简单介绍 BottomNavigationBarItem 的构造方法，各位读者可以自行实现组件效果。

```
const BottomNavigationBarItem({
  // 图标
  @required this.icon,
  // 文字
  this.title,
  // 选中图标
  Widget activeIcon,
  // 背景色
  this.backgroundColor,
})
```

5.3.4 CupertinoTabBar 和 PageView 相关组件

CupertinoTabBar 组件继承自 StatelessWidget，属于无状态组件。PageView 组件继承自 StatefulWidget，属于有状态组件。CupertinoTabBar 主要用来实现底部导航，配合 PageView 可实现导航切换页面的效果。CupertinoTabBar 组件的构造方法如下。

```
CupertinoTabBar({
    Key key,
    // 导航项
    @required this.items,
    this.onTap,
    this.currentIndex = 0,
    this.backgroundColor,
    // 选中色
    this.activeColor,
    // 未选中色
    this.inactiveColor = CupertinoColors.inactiveGray,
    this.iconSize = 30.0,
    // 边框
    this.border = const Border(
      top: BorderSide(
        color: _kDefaultTabBarBorderColor,
        width: 0.0, // One physical pixel.
        style: BorderStyle.solid,
      ),
    ),
})
```

接下来我们来看一下 PageView 组件的构造方法，如下。

```
PageView({
    Key key,
    // 滚动方向
    this.scrollDirection = Axis.horizontal,
    this.reverse = false,
    // 页面控制
    PageController controller,
    // 滚动效果
    this.physics,
    this.pageSnapping = true,
    // 通过页面改变监听
    this.onPageChanged,
    // 子组件
    List<Widget> children = const <Widget>[],
    this.dragStartBehavior = DragStartBehavior.down,
  })
```

了解了 CupertinoTabBar 与 PageView 的基本构造方法后，我们结合两者，通过代码实现底部导航切换页面的功能，并支持页面滑动切换，具体代码如下。

```
class CupertinoTabBarState extends State<CupertinoTabBarSamples> {
  // 默认选中第一项
  int _selectedIndex = 0;
  var _pageController = new PageController(initialPage: 0);

  @override
  void initState() {
    super.initState();
    _pageController.addListener(() {});
  }

  @override
  Widget build(BuildContext context) {
    return Scaffold(
      appBar: AppBar(
        title: Text('CupertinoTabBar Demo'),
      ),
      // body 主体内容用 PageView 组件实现
      body: PageView(
```

```
        // 监听控制类
        controller: _pageController,
        onPageChanged: _onItemTapped,
        children: <Widget>[
          Text('Index 0: Home'),
          Text('Index 1: Business'),
          Text('Index 2: School'),
        ],
      ),
      // 底部导航栏用 CupertinoTabBar 组件实现
      bottomNavigationBar: CupertinoTabBar(
        // 导航集合
        items: <BottomNavigationBarItem>[
          BottomNavigationBarItem(icon: Icon(Icons.home), title: Text('Home')),
          BottomNavigationBarItem(
              icon: Icon(Icons.business), title: Text('Business')),
          BottomNavigationBarItem(
              icon: Icon(Icons.school), title: Text('School')),
        ],
        currentIndex: _selectedIndex,
        onTap: setPageViewItemSelect,
      ),
    );
  }

  void _onItemTapped(int index) {
    setState(() {
      _selectedIndex = index;
    });
  }

  // 底部点击切换
  void setPageViewItemSelect(int indexSelect) {
    _pageController.animateToPage(indexSelect,
        duration: const Duration(milliseconds: 300), curve: Curves.ease);
  }
}
```

执行以上代码，效果如图 5-26 所示。

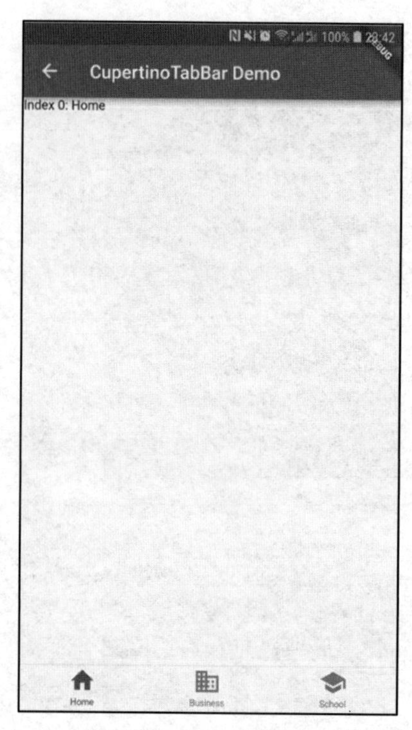

图 5-26 CupertinoTabBar 组件结合 PageView 组件效果

第 6 章

Flutter 常用组件（下）

第 5 章介绍了部分 Flutter 常用组件，本章将继续介绍 Flutter 常用组件：表单类组件、列表滚动组件、Dialog 组件、表格组件。本章的内容同样非常重要，希望各位读者能够理解，并结合案例进行实践。

6.1　表单类组件

本节将以 From 组件为例，介绍表单类组件的相关内容。Flutter 中的 Form 组件和 HTML 里的 Form 的作用和用法基本一致，主要用于提交一系列表单信息，如注册信息、登录信息等。

Form 组件继承自 StatefulWidget，是有状态组件。Form 组件的里面每一个子组件必须是 FormField 类型的，Form 组件一般也会和 TextFormField 组件搭配使用。

我们先来看一下 From 组件和 FormField 组件的构造方法及属性参数含义，具体如下。

```
const Form({
    Key key,
    // 子组件
    @required this.child,
    // 是否自动校验子组件输入的内容
    this.autovalidate = false,
    // 返回按键处理
    this.onWillPop,
    // 子组件内容发生变化时触发此回调
    this.onChanged,
  })
```

```
// FormField 组件构造方法
const FormField({
    Key key,
    @required this.builder,
    // 保存回调
    this.onSaved,
    // 验证回调
    this.validator,
    // 初始值
    this.initialValue,
    // 是否自动校验
    this.autovalidate = false,
    // 是否可用
    this.enabled = true,
  })
```

TextFormField 组件也是实现表单功能常用的组件。TextFormField 组件继承自 FormField，主要负责实现表单里的文本输入框功能，其构造方法和属性参数含义如下。

```
// 大部分属性前面都介绍过，这里不再重复赘述
TextFormField({
    Key key,
    this.controller,
    String initialValue,
    FocusNode focusNode,
    InputDecoration decoration = const InputDecoration(),
    TextInputType keyboardType,
    TextCapitalization textCapitalization = TextCapitalization.none,
    TextInputAction textInputAction,
    TextStyle style,
    TextDirection textDirection,
    TextAlign textAlign = TextAlign.start,
    bool autofocus = false,
    bool obscureText = false,
    bool autocorrect = true,
    bool autovalidate = false,
    bool maxLengthEnforced = true,
    int maxLines = 1,
    int maxLength,
    VoidCallback onEditingComplete,
```

```
    // 提交数据
    ValueChanged<String> onFieldSubmitted,
    FormFieldSetter<String> onSaved,
    FormFieldValidator<String> validator,
    List<TextInputFormatter> inputFormatters,
    bool enabled = true,
    double cursorWidth = 2.0,
    Radius cursorRadius,
    Color cursorColor,
    Brightness keyboardAppearance,
    EdgeInsets scrollPadding = const EdgeInsets.all(20.0),
    bool enableInteractiveSelection = true,
    InputCounterWidgetBuilder buildCounter,
  })
```

以上我们介绍了几个最常用的表单类组件。那么这几个组件是如何关联使用的呢？Form 和 FormField 是如何管理和通信的呢？答案就是通过 key 和 FormState，方法如下。

```
GlobalKey<FormState> _formKey = new GlobalKey<FormState>();
FormState _formState;
...

_formState = _formKey.currentState;

Form(
    key: _formKey,
...
```

通过调用 FormState 的相关方法，我们就可以用 FormState 对 Form 的子组件 FormField 进行统一操作，方法如下。

```
// 保存
_formState.save();
// 清空重置
_formState.reset();
// 验证
_formState.validate();
```

调用这几个方法后，会同时调用 FormField 组件里的对应方法，这样就实现了 Form 和 FormField 的通信、控制、管理等相关操作。

接下来我们看一个示例，主要功能是实现一个登录表单界面，界面上包含用户名输入框、密码输入框和一个登录按钮，按下登录按钮后即可提交表单内容，代码如下。

```
class FormSamplesState extends State<FormSamples> {
  GlobalKey<FormState> _formKey = new GlobalKey<FormState>();
  FormState _formState;
  String _name;
  String _password;

  @override
  void initState() {
    super.initState();
  }

  @override
  Widget build(BuildContext context) {
    return Scaffold(
      appBar: AppBar(title: Text('Form Widget'), primary: true),
      body: Form(
        key: _formKey,
        child: Column(
          children: <Widget>[
            TextFormField(
              decoration: const InputDecoration(
                  icon: Icon(Icons.person),
                  hintText: '请输入用户名',
                  labelText: '用户名',
                  prefixText: '用户名：'),
              onSaved: (String value) {
                _name = value;
              },
              validator: (String value) {
                return value.contains('@') ? '用户名里不要使用@符号' : null;
              },
            ),
            TextFormField(
              decoration: InputDecoration(
                  labelText: '密码',
                  icon: Icon(Icons.lock),
                  hintText: '请输入密码',
                  prefixText: '密码：'),
```

```
              maxLines: 1,
              maxLength: 32,
              obscureText: true,
              keyboardType: TextInputType.numberWithOptions(),
              validator: (value) {},
              onSaved: (value) {
                _password = value;
              },
              onFieldSubmitted: (value) {},
            ),

            /// 长按被 Tooltip 包裹的控件弹出 tips
            Tooltip(
              message: '表单提交',
              child: RaisedButton(
                child: Text('登录'),
                onPressed: () {
                  onSubmit();
                },
              ),
            ),
          ],
        ),
      ),
    );
  }

  void onSubmit() {
    final _formState = _formKey.currentState;
    if (_formState.validate()) {
      _formState.save();
      showDialog<void>(
          context: context,
          barrierDismissible: false,
          builder: (BuildContext context) {
            return AlertDialog(
              title: Text("提示"),
              content: Column(
                children: <Widget>[
```

```
                Text(
                  'Name: $_name',
                  style: TextStyle(
                      fontSize: 18, decoration: TextDecoration.none),
                ),
                Text(
                  'Password: $_password',
                  style: TextStyle(
                      fontSize: 18, decoration: TextDecoration.none),
                ),
              ],
            ),
            actions: <Widget>[
              new FlatButton(
                  onPressed: () => Navigator.of(context).pop(false),
                  child: new Text("取消")),
              new FlatButton(
                  onPressed: () {
                    Navigator.of(context).pop(true);
                  },
                  child: new Text("确定")),
            ],
          );
        });
  }
}
}
```

执行以上代码，效果如图 6-1 所示。

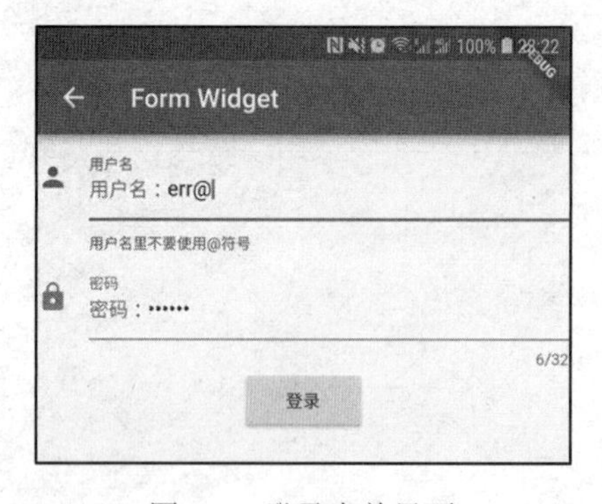

图 6-1　登录表单界面

6.2　列表滚动组件

在实际开发中，经常会涉及列表滚动功能。本节我们将对 Flutter 中的列表滚动组件进行详细分析。Flutter 列表滚动组件主要有 CustomScrollView 组件、ListView 组件、GridView 组件、ScrollView 组件、ExpansionPanel 组件等。

6.2.1　CustomScrollView 组件

CustomScrollView 组件是一个自定义的内容滚动组件，它的内部通过 Flutter 的 Sliver 来组装、包裹滚动组件内容，其中可以放置任何实现滚动功能的组件，CustomScrollView 组件也是一个使用非常频繁的组件，其继承关系如下。

```
CustomScrollView -> ScrollView -> StatelessWidget
```

CustomScrollView 是一个无状态组件，继承自 ScrollView，也扩展了 ScrollView 的功能。其最大的特点是，内部包裹的滚动组件（如 SliverAppBar、SliverGrid 等）都具有 Sliver 特性。

CustomScrollView 功能非常强大，我们可以将其与其他的 Sliver 组件拼装使用，实现更复杂的效果。CustomScrollView 组件的构造方法和属性参数含义如下。

```
const CustomScrollView({
    Key key,
    // 滚动方向
    Axis scrollDirection = Axis.vertical,
    // 是否反向显示
    bool reverse = false,
    // 滚动控制对象
    ScrollController controller,
    // 是否是与父级关联的主滚动视图
    bool primary,
    // 滚动视图应如何响应用户输入
    ScrollPhysics physics,
    // 是否根据正在查看的内容确定滚动视图范围
    bool shrinkWrap = false,
    Key center,
    double anchor = 0.0,
    // 缓存区域
    double cacheExtent,
    // 内部 Sliver 组件
```

```
  this.slivers = const <Widget>[],
  int semanticChildCount,
  DragStartBehavior dragStartBehavior = DragStartBehavior.down,
})
```

CustomScrollView 的可配置拓展功能很多。接下来我们通过一个示例来看一下如何实现 CustomScrollView 组件，代码如下。

```
@override
 Widget build(BuildContext context) {
   return Material(
     child: CustomScrollView(
       // slivers 里面放置 Sliver 滚动组件
       slivers: <Widget>[
         // 放置一个顶部标题栏
         SliverAppBar(
           // 指明是否固定在顶部
           pinned: true,
           // 展开高度
           expandedHeight: 250.0,
           // 可展开区域
           flexibleSpace: FlexibleSpaceBar(
             title: const Text('CustomScrollView'),
             background: Image.asset("assets/image_appbar.jpg",fit: BoxFit.cover,),
           ),
         ),
         // 放置一个 SliverGrid 组件
         SliverGrid(
           // 设置 Grid 属性
           // SliverGridDelegateWithMaxCrossAxisExtent:
           // 按照设置最大扩展宽度计算 item 个数
           // SliverGridDelegateWithFixedCrossAxisCount:
           // 设置每行 item 个数
           gridDelegate: SliverGridDelegateWithMaxCrossAxisExtent(
             // item 最大宽度
             maxCrossAxisExtent: 200.0,
             // 主轴 item 间隔
             mainAxisSpacing: 10.0,
             // 交叉轴 item 间隔
             crossAxisSpacing: 10.0,
```

```
          // item 宽高比
          childAspectRatio: 4.0,
        ),
        // 设置 item 的布局及属性
        // SliverChildListDelegate: 适用于有固定数量 item 的 List
        // SliverChildBuilderDelegate:适用于无固定数量 item 的 List
        delegate: SliverChildBuilderDelegate(
          (BuildContext context, int index) {
            return Container(
              alignment: Alignment.center,
              color: Colors.teal[100 * (index % 9)],
              child: Text('grid item $index'),
            );
          },
          // 20 个 item
          childCount: 20,
        ),
      ),
      // 指定 item 高度
      SliverFixedExtentList(
        // item 固定高度
        itemExtent: 50.0,
        // 设置 item 布局和属性
        delegate: SliverChildBuilderDelegate(
          (BuildContext context, int index) {
            return Container(
              alignment: Alignment.center,
              color: Colors.lightBlue[100 * (index % 9)],
              child: Text('list item $index'),
            );
          },
          childCount: 20,
        ),
      ),
    ],
  ),
);
}
```

执行以上代码，效果如图 6-2 所示。

图 6-2　CustomScrollView 组件效果

6.2.2　ListView 组件

ListView 组件主要用于实现线性列表布局。ListView 的继承关系如下。

```
ListView -> BoxScrollView -> ScrollView
```

ListView 也继承自 ScrollView 组件，扩展了 ScrollView 的特点。ListView 一般用来实现列表类的效果，其构造方法和属性参数含义如下。

```
ListView({
    Key key,
    // 滚动排列方向
    Axis scrollDirection = Axis.vertical,
    bool reverse = false,
    ScrollController controller,
    bool primary,
    // 物理滑动响应动画
    ScrollPhysics physics,
```

```
    // 是否根据子组件的总高度/长度设置 ListView 的长度，默认值为 false
    bool shrinkWrap = false,
    EdgeInsetsGeometry padding,
    // item 固定高度
    this.itemExtent,
    // 是否将 item 包裹在 AutomaticKeepAlive 组件中
    bool addAutomaticKeepAlives = true,
    // 是否将 item 包裹在 RepaintBoundary 组件中
    bool addRepaintBoundaries = true,
    bool addSemanticIndexes = true,
    double cacheExtent,
    // 子 item 元素
    List<Widget> children = const <Widget>[],
    int semanticChildCount,
    DragStartBehavior dragStartBehavior = DragStartBehavior.down,
  })
```

在了解了 ListView 组件的构造方法和属性参数含义后，我们使用 ListView 组件来实现一个类似于通讯录联系人列表的示例，代码如下。

```
// ListView 最简单的用法
ListView(
  shrinkWrap: true,
  padding: const EdgeInsets.all(20.0),
  children: <Widget>[
    const Text('I\'m dedicating every day to you'),
    const Text('Domestic life was never quite my style')
  ],
),

// 复杂用法
// 定义一个 List
List<String> items = <String>[
  'A',
  'B',
  'C',
  … …
  'M',
  'N',
  'O',
```

```
];

// 定义一个枚举类型来设置 item 显示几行
enum _MaterialListType {
  oneLine,
  twoLine,
  threeLine,
}

class ListViewSamples extends StatefulWidget {
  @override
  State<StatefulWidget> createState() {
    return ListViewSamplesState();
  }
}

class ListViewSamplesState extends State<ListViewSamples> {
  List widgets = [];

  @override
  void initState() {
    super.initState();
  }

  @override
  Widget build(BuildContext context) {
    return Scaffold(
      appBar: AppBar(
        title: Text('ListView'),
      ),
      body: listView4(),
    );
  }

  // 最简单的 ListView
  Widget listView1() {
    return ListView(
      children: <Widget>[
        Text(
```

```
          'data',
          style: TextStyle(fontSize: 30),
        ),
        Text(
          'data',
          style: TextStyle(fontSize: 30),
        )
      ],
    );
}

// 动态封装 ListView，使用 ListTile 作为 item
Widget listView2() {
  // listTiles 为 item 布局集合
  Iterable<Widget> listTiles = items.map<Widget>((String string) {
    return getItem(string);
  });
  ListTile.divideTiles(context: context, tiles: listTiles);
  return ListView(
    children: listTiles.toList(),
  );
}

// 使用 ListView.builder 构造
Widget listView3() {
  // item 组件集合
  return ListView.builder(
    // 设置 item 数量
    itemCount: items.length,
    itemBuilder: (BuildContext context, int position) {
      return getItem(items.elementAt(position));
    },
  );
}

// 通过 ListView.custom 构建 ListView
Widget listView4() {
  // SliverChildListDelegate：适用于有固定数量 item 的 List
  // SliverChildBuilderDelegate:适用于无固定数量 item 的 List
```

```
  return ListView.custom(
    // 设置 item 构建属性
    childrenDelegate:
        SliverChildBuilderDelegate((BuildContext context, int index) {
      // 返回 item 布局
      return ListTile(
        isThreeLine: true,
        dense: true,
        leading: ExcludeSemantics(
            child: CircleAvatar(child: Text(items.elementAt(index)))),
        title: Text('This item represents .'),
        subtitle: Text("$index"),
        trailing: Icon(Icons.info, color: Theme.of(context).disabledColor),
      );
    }, childCount: 13),
  );
}

/// 通过 ListView.separated 构建 ListView
Widget listView5() {
  // 有分隔线
  return ListView.separated(
    // item 数量
    itemCount: items.length,
    // 分隔线属性
    separatorBuilder: (BuildContext context, int index) {
      return Container(height: 1, color: Colors.pink);
    },
    // 构建 item 布局
    itemBuilder: (BuildContext context, int index) {
      return ListTile(
        isThreeLine: true,
        dense: true,
        leading: ExcludeSemantics(
            child: CircleAvatar(child: Text(items.elementAt(index)))),
        title: Text('This item represents .'),
        subtitle: Text(items.elementAt(index)),
        trailing: Icon(Icons.info, color: Theme.of(context).disabledColor),
      );
```

```
      },
    );
  }

  // 用来获取 item 布局的方法
  Widget getItem(String item) {
    // if (i.isOdd) {
    //   return Divider();
    // }
    return GestureDetector(
      child: Padding(
        padding: EdgeInsets.all(10.0),
        child: ListTile(
            dense: true,
            title: Text('Two-line ' + item),
            trailing: Radio<_MaterialListType>(
              value: _MaterialListType.twoLine,
              groupValue: _MaterialListType.twoLine,
              onChanged: changeItemType,
            )),
      ),
      onTap: () {
        setState(() {
          // print('row $i');
        });
      },
      onLongPress: () {},
    );
  }

  void changeItemType(_MaterialListType type) {
    print("changeItemType");
  }
}
```

执行以上代码，效果如图 6-3 所示。

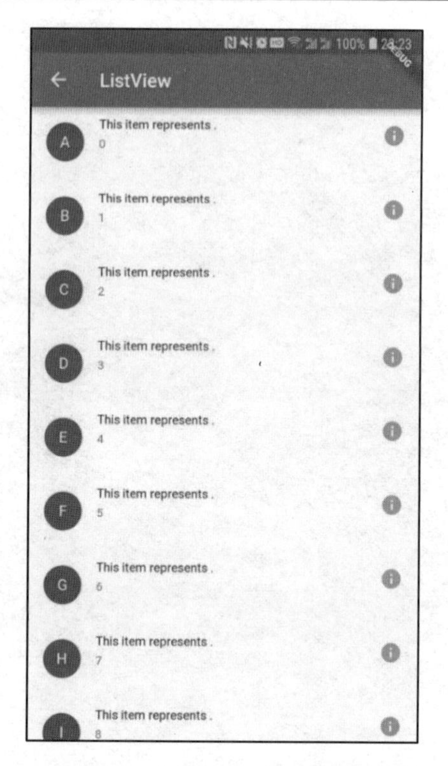

图 6-3　ListView 组件效果

6.2.3　GridView 组件

GridView 组件的用法和 ListView 组件的用法类似，各位读者可以对比着学习。GridView 组件主要用于实现网格列表的排列效果。GridView 的继承关系如下。

```
GridView -> BoxScrollView -> ScrollView
```

GridView 也继承自 ScrollView 组件，扩展了 ScrollView 的特点，其构造方法和属性参数含义如下。

```
GridView({
    Key key,
    Axis scrollDirection = Axis.vertical,
    bool reverse = false,
    ScrollController controller,
    bool primary,
    ScrollPhysics physics,
    bool shrinkWrap = false,
```

```
    EdgeInsetsGeometry padding,
    // 控制 GridView 组件的 item 如何排列
    @required this.gridDelegate,
    bool addAutomaticKeepAlives = true,
    bool addRepaintBoundaries = true,
    bool addSemanticIndexes = true,
    double cacheExtent,
    // item 集合
    List<Widget> children = const <Widget>[],
    int semanticChildCount,
  })
```

GridView 组件的很多属性和前面介绍的 CustomScrollView 和 ListView 一致，这里就不重复讲解了。有了前面学习 ListView 组件的基础，掌握 GridView 组件用法就不在话下了，我们直接来看一个示例，功能是实现一个网格列表布局，在本例中我们将给出多种构造 GridView 组件的方式，代码如下。

```
// GridView 的简单用法
body: GridView.count(
        primary: false,
        padding: const EdgeInsets.all(20.0),
        crossAxisSpacing: 10.0,
        // 每行多少个 item
        crossAxisCount: 2,
        children: <Widget>[
          const Text('He\'d have you all unravel at the'),
          const Text('Heed not the rabble'),
        ],
      ),

// 稍微复杂的用法
class GridViewSamplesState extends State<GridViewSamples> {
  @override
  void initState() {
    super.initState();
  }

  @override
  Widget build(BuildContext context) {
    return Scaffold(
```

```
      appBar: AppBar(
        title: Text('GridView'),
        backgroundColor: Colors.teal,
        primary: true,
      ),
      body: gridView1(),
    );
  }
// 构造 GriView 的方式 1
Widget gridView1() {
  return GridView(
    // 设置 Grid 属性
    // SliverGridDelegateWithMaxCrossAxisExtent:
    // 按照设置最大扩展宽度计算 item 个数
    // SliverGridDelegateWithFixedCrossAxisCount:
    // 设置每行 item 个数
    gridDelegate: SliverGridDelegateWithFixedCrossAxisCount(
      crossAxisCount: 2,
      crossAxisSpacing: 10,
      mainAxisSpacing: 10,
    ),
    children: <Widget>[
      Image.asset(
        'assets/image_appbar.jpg',
        fit: BoxFit.cover,
      ),
      Image.asset(
        'assets/image_appbar.jpg',
        fit: BoxFit.cover,
      ),
    ],
  );
}
// 构造 GriView 的方式 2
Widget gridView2() {
  return GridView.builder(
    // item 总数
    itemCount: 20,
    // 构建 item
```

```
    itemBuilder: (BuildContext context, int index) {
      // 构建带有头部、底部、中间内容的 item
      return GridTile(
        header: GridTileBar(
          title: Text(
            'header',
            style: TextStyle(fontSize: 20),
          ),
          backgroundColor: Colors.black45,
          leading: Icon(
            Icons.star,
            color: Colors.white,
          ),
        ),
        child: Image.asset('assets/image_appbar.jpg'),
        footer: GridTileBar(
          title: Text(
            'bottom',
            style: TextStyle(fontSize: 20),
          ),
          backgroundColor: Colors.black45,
        ),
      );
    },
    // 设置 GridView 排列属性
    gridDelegate: SliverGridDelegateWithFixedCrossAxisCount(
      crossAxisCount: 2,
      crossAxisSpacing: 10,
      mainAxisSpacing: 10,
    ),
  );
}
// 构造 GriView 的方式 3
Widget gridView3() {
  return GridView.custom(
    // 设置 GridView 属性
    gridDelegate: SliverGridDelegateWithFixedCrossAxisCount(
      crossAxisCount: 2,
      crossAxisSpacing: 10,
```

```
      mainAxisSpacing: 10,
      childAspectRatio: 2,
    ),
    // 设置 item 属性
    childrenDelegate:
        SliverChildBuilderDelegate((BuildContext context, int index) {
      return Container(
        child: Text(
          'GridTile',
          style: TextStyle(fontSize: 16),
        ),
      );
    }, childCount: 20),
  );
}
// 构造 GriView 的方式 4
Widget gridView4() {
  return GridView.count(
    crossAxisCount: 2,
    mainAxisSpacing: 10,
    crossAxisSpacing: 10,
    childAspectRatio: 1,
    children: <Widget>[
      GridTile(
        child: Image.asset('assets/image_appbar.jpg'),
      ),
      GridTile(
        child: Image.asset('assets/image_appbar.jpg'),
      )
    ],
  );
}

// 构造 GriView 的方式 5
// 通过 GridView.extent 构造 GridView，根据最大宽度自动计算 item 数量
Widget gridView5() {
  return GridView.extent(
    // 最大宽度
    maxCrossAxisExtent: 150,
```

```
crossAxisSpacing: 10,
mainAxisSpacing: 10,
childAspectRatio: 1,
children: <Widget>[
  GridTile(
    header: GridTileBar(
      title: Text(
        'header',
        style: TextStyle(fontSize: 20),
      ),
      backgroundColor: Colors.black45,
      leading: Icon(
        Icons.star,
        color: Colors.white,
      ),
    ),
    child: Image.asset('assets/image_appbar.jpg'),
    footer: GridTileBar(
      title: Text(
        'bottom',
        style: TextStyle(fontSize: 20),
      ),
      backgroundColor: Colors.black45,
    ),
  ),
  GridTile(
    header: GridTileBar(
      title: Text(
        'header',
        style: TextStyle(fontSize: 20),
      ),
      backgroundColor: Colors.black45,
      leading: Icon(
        Icons.star,
        color: Colors.white,
      ),
    ),
    child: Image.asset('assets/image_appbar.jpg'),
    footer: GridTileBar(
```

```
          title: Text(
            'bottom',
            style: TextStyle(fontSize: 20),
          ),
          backgroundColor: Colors.black45,
        ),
      ),
    ],
  );
 }
}
```

执行以上代码，效果如图 6-4 所示。

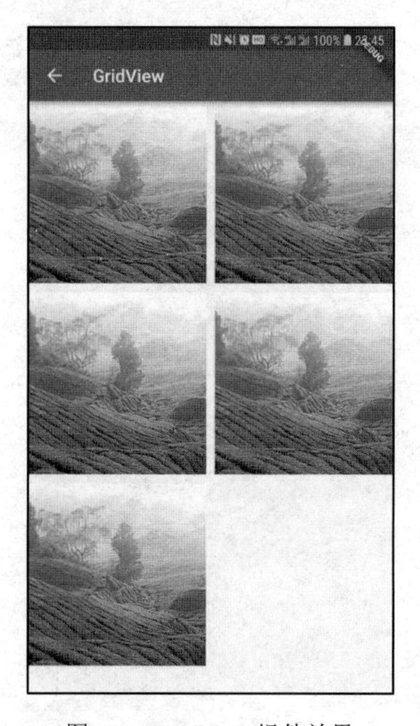

图 6-4　GridView 组件效果

GridView 组件一般用来实现网格化布局页面，例如电商网站商品展示列表页等。大家可以模仿以上示例编写一些其他的网格化布局页面，加深对 GridView 组件用法的理解。

6.2.4　ScrollView 组件

ScrollView 组件是父类组件，一般不单独使用，我们通常使用它的扩展类组件进行开发，如 CustomScrollView、ListView、GridView 等。

ScrollView 组件可用于进行滚动条样式的设置，一般通过滚动类组件实现滚动页面后，我们希望页面侧边显示一个滚动条，这种情况下可以使用 Scrollbar 来实现，代码如下。

```
Widget scroll1() {
  return Scrollbar(
    child: ListView.separated(
      itemCount: 20,
      separatorBuilder: (BuildContext context, int index) {
        return Container(height: 1, color: Colors.black87);
      },
      itemBuilder: (BuildContext context, int index) {
        return ListTile(
          isThreeLine: true,
          dense: true,
          leading:
              ExcludeSemantics(child: CircleAvatar(child: Text('leading'))),
          title: Text('This item represents .'),
          subtitle: Text('subtitle'),
          trailing: Icon(Icons.info, color: Theme.of(context).disabledColor),
        );
      },
    ),
  );
}
```

下面我们来看一个内部只能包裹一个子组件的 ScrollView 拓展组件 SingleChildScrollView，具体使用方法如下。

```
Widget scroll2() {
  // 只能包裹一个组件
  return SingleChildScrollView(
    child: Column(
      children: <Widget>[
        Container(
          color: Colors.yellow,
          height: 620.0,
        ),
        Container(
          color: Colors.orange,
```

```
        height: 720.0,
      ),
    ],
  ),
 );
}
```

有时我们需要自定义内部子组件排列方式，这时可以使用 CustomMultiChildLayout 组件或 CustomSingleChildLayout 组件，这两个组件不常用，大家了解即可，此处不进行详细说明。

6.2.5 ExpansionPanel 组件

ExpansionPanel 组件主要用来实现类似于 QQ 分组的功能，一般搭配 ExpansionPanelList 组件一起使用。ExpansionPanelList 组件继承自 StatefulWidget，属于有状态组件，其构造方法和属性参数含义如下。

```
const ExpansionPanelList({
    Key key,
    // 子组件 ExpansionPanel 集合
    this.children = const <ExpansionPanel>[],
    // 控制展开或关闭回调
    this.expansionCallback,
    // 展开动画执行时长
    this.animationDuration = kThemeAnimationDuration,
  })
```

ExpansionPanel 组件的构造方法和属性参数含义如下。

```
ExpansionPanel({
    // 标题构造器
    @required this.headerBuilder,
    // 内容区域
    @required this.body,
    // 指明是否展开
    this.isExpanded = false
  })
```

本节我们不给出 ExpansionPanel 组件和 ExpansionPanelList 组件配合使用的示例代码，希望各位读者自己动手，编写一个能够实现分组功能的应用。

6.3 Dialog 组件

Dialog 组件也比较常用，主要用于实现对话框效果。Flutter 中 Dialog 组件一般分为 AlertDialog 组件（实现警告提示对话框）、SimpleDialog 组件（实现有列表选项的对话框）、CupertinoAlertDialog 组件（实现警告提示对话框，是 iOS 风格的 AlertDialog）、CupertinoDialog 组件（实现 iOS 风格对话框）几种，效果分别如图 6-5、图 6-6、图 6-7、图 6-8 所示。

图 6-5　AlertDialog 组件效果

图 6-6　SimpleDialog 组件效果

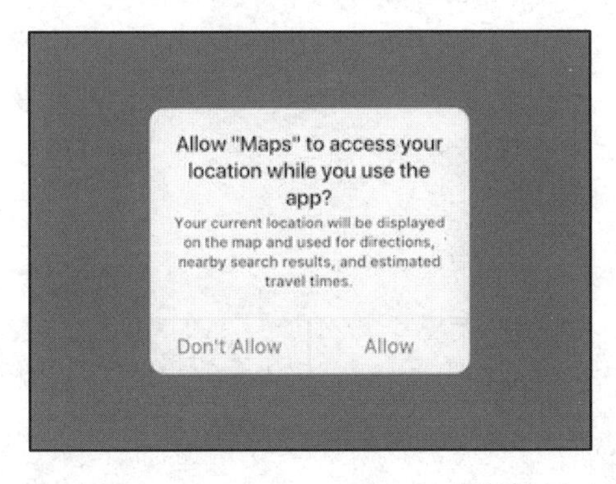

图 6-7　CupertinoAlertDialog 组件效果

图 6-8　CupertinoDialog 组件效果

在 Flutter 中，弹出对话框的方法有以下几种。

- showDialog：弹出 Material 风格对话框。
- showCupertinoDialog：弹出 iOS 风格对话框。

- showGeneralDialog：弹出自定义对话框，默认状态下，点击空白处弹出的窗口不消失。
- showAboutDialog：弹出应用中与页面相关的对话框。

了解了实现对话框的组件及 Flutter 中弹出对话框的方法后，我们来实现一个对话框示例，此处以 AlertDialog 组件为例，具体代码如下。

```
enum Option { A, B, C }
enum Location { Barbados, Bahamas, Bermuda }

// AlertDialog
 Future<void> dialog1(BuildContext context) async {
   return showDialog<void>(
       context: context,
       // 指明点击周围空白区域对话框是否消失
       barrierDismissible: false,
       builder: (BuildContext context) {
         return AlertDialog(
           title: Text("提示"),
           content: new Text("是否退出"),
           actions: <Widget>[
             new FlatButton(
                 onPressed: () => Navigator.of(context).pop(false),
                 child: new Text("取消")),
             new FlatButton(
                 onPressed: () {
                   Navigator.of(context).pop(true);
                 },
                 child: new Text("确定"))
           ],
         );
       });
 }

// AlertDialog
 Future<void> officialDialog2(BuildContext context) async {
   return showDialog<void>(
     context: context,
     barrierDismissible: false,
     builder: (BuildContext context) {
```

```
      return AlertDialog(
        title: Text('Rewind and remember'),
        content: SingleChildScrollView(
          child: ListBody(
            children: <Widget>[
              Text('You will never be satisfied.'),
              Text('You\'re like me. I'm never satisfied.'),
            ],
          ),
        ),
        actions: <Widget>[
          FlatButton(
            child: Text('Regret'),
            onPressed: () {
              Navigator.of(context).pop();
            },
          ),
        ],
      );
    },
  );
}

// SimpleDialog
Future<void> dialog3(BuildContext context) async {
  return showDialog<void>(
      context: context,
      barrierDismissible: false,
      builder: (BuildContext context) {
        return SimpleDialog(
          title: Text("提示"),
        );
      });
}

// SimpleDialog
Future<void> dialog4(BuildContext context) async {
  switch (await showDialog<Option>(
      context: context,
```

```
    builder: (BuildContext context) {
      return SimpleDialog(
        title: const Text('Select Answer'),
        children: <Widget>[
          SimpleDialogOption(
            onPressed: () {
              Navigator.pop(context, Option.A);
            },
            child: const Text('A'),
          ),
          SimpleDialogOption(
            onPressed: () {
              Navigator.pop(context, Option.B);
            },
            child: const Text('B'),
          ),
          SimpleDialogOption(
            onPressed: () {
              Navigator.pop(context, Option.C);
            },
            child: const Text('C'),
          ),
        ],
      );
    })) {
  case Option.A:
    // Let's go.
    // ...
    print('A');
    break;
  case Option.B:
    // ...
    print('B');
    break;
  case Option.C:
    // ...
    print('C');
    break;
```

```
    }
}
```

执行以上代码，效果如图 6-9 所示。

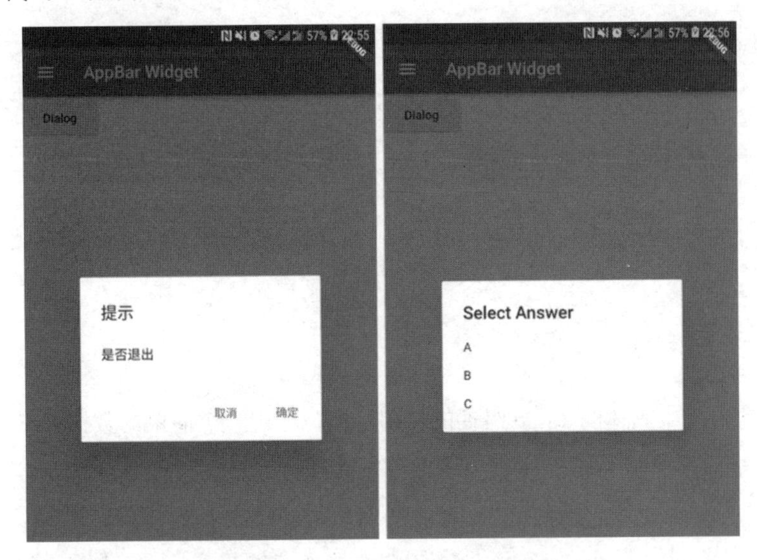

图 6-9　通过 AlertDialog 组件实现的对话框

一般情况下，Flutter 提供的 Dialog 组件基本可以满足大部分的开发需求，希望大家可以多多练习。

6.4　表格组件

本节主要讲解 Flutter 中表格组件的用法。在 Flutter 中，表格主要通过 Table 组件和 DataTable 组件来完成绘制。除介绍以上两种组件外，本节还会介绍 PaginatedDataTable 组件。

6.4.1　Table 组件

Table 组件的继承关系是 Table -> RenderObjectWidget -> Widget。通过 Table 组件绘制表格时，表格中的行用 TableRow 控制，列用 columnWidths 控制。Table 组件的构造方法和属性参数含义如下。

```
Table({
    Key key,
    // 每行的 TableRow 集合
```

```
    this.children = const <TableRow>[],
    // 设置每列的宽度
    this.columnWidths,
    // 每列宽度默认情况下相等
    this.defaultColumnWidth = const FlexColumnWidth(1.0),
    // 文字方向
    this.textDirection,
    // 设置表格边框
    this.border,
    // 默认垂直方向对齐
    this.defaultVerticalAlignment = TableCellVerticalAlignment.top,
    // defaultVerticalAlignment 为 baseline 时，配置生效
    this.textBaseline,
  })
```

下面我们来看一个具体的 Table 组件的使用示例，实现一个简单的数据表格，代码如下。

```
class TableSamplesState extends State<TableSamples> {
  @override
  void initState() {
    super.initState();
  }

  @override
  Widget build(BuildContext context) {
    return Scaffold(
        appBar: AppBar(title: Text('Table Demo'), primary: true),
        body: Padding(
          padding: EdgeInsets.all(10),
          // 使用 Table 绘制表格
          child: Table(
            // 有多种设置宽度的方式
            columnWidths: {
              // 固定列宽度
              0: FixedColumnWidth(50),
              /// 弹性列宽度
              1: FlexColumnWidth(1),
              /// 宽度占所在容器的百分比（0~1）
              2: FractionColumnWidth(0.5),
              3: IntrinsicColumnWidth(flex: 0.2),
              4: MaxColumnWidth(
```

```
        FixedColumnWidth(100.0), FractionColumnWidth(0.1)),

    // 宽度大于容器的 10%，小于等于 100px
    5: MinColumnWidth(
        FixedColumnWidth(100.0), FractionColumnWidth(0.1)),
  },
  // 设置表格边框
  border: TableBorder.all(color: Colors.black, width: 1),
  children: <TableRow>[
    // 每行内容设置
    TableRow(children: <Widget>[
      // 每个表格单元
      TableCell(
        verticalAlignment: TableCellVerticalAlignment.middle,
        child: Center(
          child: Text(
            'Title1',
            style:
                TextStyle(fontSize: 18, fontWeight: FontWeight.bold),
          ),
        ),
      ),
      TableCell(
        verticalAlignment: TableCellVerticalAlignment.middle,
        child: Center(
          child: Text(
            'Title2',
            style:
              TextStyle(fontSize: 18, fontWeight: FontWeight.bold),
          ),
        ),
      ),
      TableCell(
        verticalAlignment: TableCellVerticalAlignment.middle,
        child: Center(
          child: Text(
            'Title3',
            style:
              TextStyle(fontSize: 18, fontWeight: FontWeight.bold),
```

```
                    ),
                  ),
                ),
              ]),
              TableRow(children: <Widget>[
                TableCell(
                  child: Text('data1'),
                ),
                TableCell(
                  child: Text('data2'),
                ),
                TableCell(
                  child: Text('data3'),
                ),
              ]),
              … …
              TableRow(children: <Widget>[
                TableCell(
                  child: Text('data1'),
                ),
                TableCell(
                  child: Text('data2'),
                ),
                TableCell(
                  child: Text('data3'),
                ),
              ]),
            ],
          ),
        ));
  }
}
```

执行以上代码，效果如图 6-10 所示。

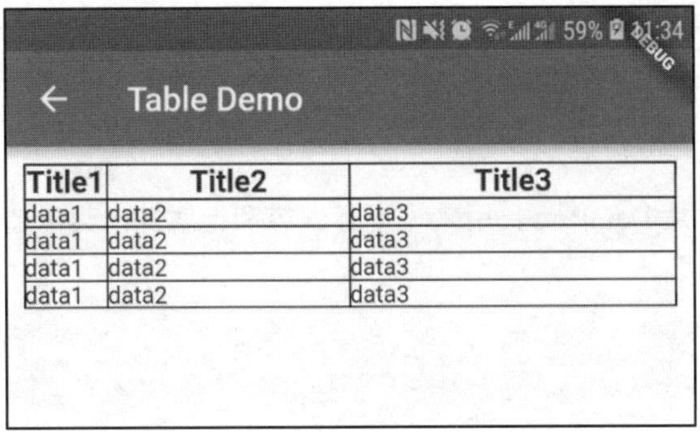

图 6-10　通过 Table 组件实现的数据表格

6.4.2　DataTable 组件

DataTable 组件也用于绘制表格，但它功能更加强大，因此也更加常用。DataTable 继承自 StatelessWidget，是一个无状态组件，其构造方法和属性参数含义如下。

```
DataTable({
    Key key,
    // 设置表头
    @required this.columns,
    // 列排序索引
    this.sortColumnIndex,
    // 是否升序排序，默认为升序
    this.sortAscending = true,
    // 点击全选
    this.onSelectAll,
    // 表格每行内容
    @required this.rows,
  })
```

DataTable 的构造方法很简单，并没有很多复杂的属性参数。在 DataTable 中，columns 属性里放置的是 DataColumn，主要用来设置列属性。DataColumn 的构造方法和属性参数含义如下。

```
const DataColumn({
    // 标签
    @required this.label,
    // 工具提示
    this.tooltip,
    // 是否包含数字
```

```
    this.numeric = false,
    // 排序时调用
    this.onSort,
  })
```

rows 属性里放置的是 DataRow，用来设置行内容和行属性，其构造方法和属性参数含义如下。

```
const DataRow({
    this.key,
    // 是否选中
    this.selected = false,
    // 选中状态改变时回调
    this.onSelectChanged,
    // 子表格单元
    @required this.cells,
  })
```

表格中每个子表格单元的内容和属性都是使用 DataCell 来设置的，我们再看一下 DataCell 的构造方法和属性参数含义，如下。

```
const DataCell(
    // 子组件，一般为 Text 或 DropdownButton
    this.child, {
    // 是否为占位符，若子组件为 Text，显示占位符文本样式
    this.placeholder = false,
    // 是否显示编辑图标，配合 onTab 使用
    this.showEditIcon = false,
    // 点击回调
    this.onTap,
  })
```

以上我们介绍了 Flutter 中实现表格的几个基础核心组件。接下来我们通过一个示例来看一下 DataTable 的具体使用方法，实现一个比较美观的数据表格，代码如下。

```
class DataTableState extends State<DataTableSamples> {
  @override
  void initState() {
    super.initState();
  }

  @override
```

```
  Widget build(BuildContext context) {
    return Scaffold(
        appBar: AppBar(title: Text('DataTable Demo'), primary: true),
        body: DataTable(
          // 行
          rows: <DataRow>[
            // 每行内容设置
            DataRow(
              cells: <DataCell>[
                // 子表格单元
                DataCell(Text('data1'), onTap: onTap),
                DataCell(Text('data2'), onTap: onTap),
                DataCell(Text('data3'), onTap: onTap),
              ],
            ),
            … …
            DataRow(
              cells: <DataCell>[
                DataCell(Text('data1'), onTap: onTap),
                DataCell(Text('data2'), onTap: onTap),
                DataCell(Text('data3'), onTap: onTap),
              ],
            )
          ],
          // 列
          columns: <DataColumn>[
            DataColumn(label: Text('DataColumn1')),
            DataColumn(label: Text('DataColumn2')),
            DataColumn(label: Text('DataColumn3')),
          ],
        ));
  }
}

void onTap() {
  print('data onTap');
}
```

执行以上代码，效果如图 6-11 所示。

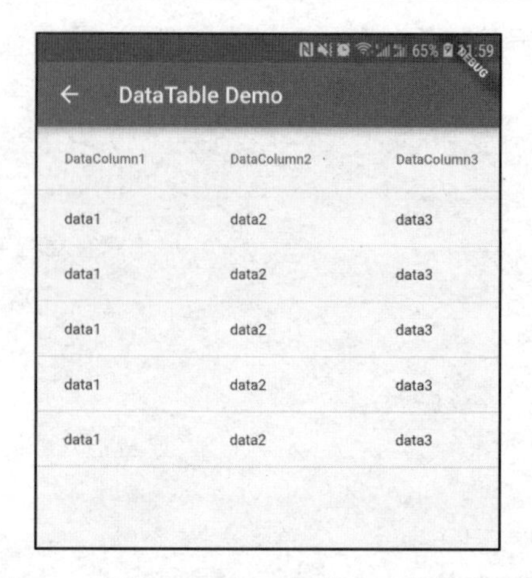

图 6-11　通过 DataTable 组件实现的数据表格

6.4.3　PaginatedDataTable 组件

PaginatedDataTable 组件属于 DataTable 组件的一种，主要用来绘制有分页类型的表格。PaginatedDataTable 继承自 StatefulWidget，是一个有状态组件。PaginatedDataTable 的构造方法和参数属性含义如下。

```
PaginatedDataTable({
    Key key,
    // 表名
    @required this.header,
    // 动作
    this.actions,
    // 列集合
    @required this.columns,
    // 列排序索引
    this.sortColumnIndex,
    // 是否升序排序
    this.sortAscending = true,
    // 点击全选回调
    this.onSelectAll,
    // 初始索引
    this.initialFirstRowIndex = 0,
    // 页码更改监听，翻页时回调
    this.onPageChanged,
    // 每页显示的行数，默认为10
```

```
    this.rowsPerPage = defaultRowsPerPage,
    // 可选择页数
    this.availableRowsPerPage = const <int>[defaultRowsPerPage,
    defaultRowsPerPage * 2, defaultRowsPerPage * 5, defaultRowsPerPage * 10],
    // 点击可选择页数下拉监听
    this.onRowsPerPageChanged,
    this.dragStartBehavior = DragStartBehavior.down,
    // 表格数据源
    @required this.source
  })
```

在使用 PaginatedDataTable 组件时，必须确保 PaginatedDataTable 的外层元素是 ListView 或 ScrollView 这种可滚动组件，并且要设置表格数据源组件 DataTableSource 提供表格数据。

下面我们通过一个示例来看一下 PaginatedDataTable 组件的具体使用方法，实现一个带有分页功能的表格，代码如下。

```
class PaginatedDataTableState extends State<PaginatedDataTableSamples> {
  TableDataSource _dataSource = TableDataSource();
  int _defalutRowPageCount = 8;
  int _sortColumnIndex;
  bool _sortAscending = true;

  @override
  void initState() {
    super.initState();
  }

  @override
  Widget build(BuildContext context) {
    return Scaffold(
        appBar: AppBar(title: Text('PaginatedDataTable Demo'), primary: true),
        // 外层用 ListView 包裹
        body: ListView(
          padding: EdgeInsets.all(10),
          children: <Widget>[
            // PaginatedDataTable
            PaginatedDataTable(
              // 表格数据源
              source: _dataSource,
```

```
        // 默认为 0
        initialFirstRowIndex: 0,
        // 全选操作
        onSelectAll: (bool checked) {
          _dataSource.selectAll(checked);
        },
        // 每页显示的行数
        rowsPerPage: _defalutRowPageCount,
        // 每页显示数量改变后回调
        onRowsPerPageChanged: (value) {
          setState(() {
            _defalutRowPageCount = value;
          });
        },
        // 设置每页可以显示的行数值
        availableRowsPerPage: [5, 8],
        // 翻页操作回调
        onPageChanged: (value) {
          print('$value');
        },
        // 是否升序排序
        sortAscending: _sortAscending,
        sortColumnIndex: _sortColumnIndex,
        // 表格头部
        header: Text('Data Header'),
        // 列
        columns: <DataColumn>[
          DataColumn(label: Text('名字')),
          DataColumn(
              label: Text('价格'),
              // 加入排序操作
              onSort: (int columnIndex, bool ascending) {
                _sort<num>((Shop p) => p.price, columnIndex, ascending);
              }),
          DataColumn(label: Text('类型')),
        ],
      ),
    ],
));
```

```
  }

  // 排序关联_sortColumnIndex 和_sortAscending
  void  sort<T>(Comparable<T> getField(Shop s), int index, bool b) {
    _dataSource._sort(getField, b);
    setState(() {
      this._sortColumnIndex = index;
      this._sortAscending = b;
    });
  }
}

class Shop {
  final String name;
  final int price;
  final String type;

  // 默认为未选中
  bool selected = false;
  Shop(this.name, this.price, this.type);
}

class TableDataSource extends DataTableSource {
  final List<Shop> shops = <Shop>[
    Shop('name', 100, '家电'),
    Shop('name2', 130, '手机'),
    Shop('三星', 130, '手机'),
    Shop('海信', 100, '家电'),
  ];
  int _selectedCount = 0;

  // 根据位置获取内容行
  @override
  DataRow getRow(int index) {
    Shop shop = shops.elementAt(index);
    return DataRow.byIndex(
        cells: <DataCell>[
          DataCell(
            Text('${shop.name}'),
```

```
          placeholder: true,
        ),
        DataCell(Text('${shop.price}'), showEditIcon: true),
        DataCell(Text('${shop.type}'), showEditIcon: false),
      ],
      selected: shop.selected,
      index: index,
      onSelectChanged: (bool isSelected) {
        if (shop.selected != isSelected) {
          _selectedCount += isSelected ? 1 : -1;
          shop.selected = isSelected;
          notifyListeners();
        }
      });
}

@override

// 行数是否不确定
bool get isRowCountApproximate => false;

@override

// 行数
int get rowCount => shops.length;

@override

// 选中的行数
int get selectedRowCount => _selectedCount;

void selectAll(bool checked) {
  for (Shop shop in shops) {
    shop.selected = checked;
  }
  _selectedCount = checked ? shops.length : 0;
  notifyListeners();
}
```

```
  // 排序,
  void _sort<T>(Comparable<T> getField(Shop shop), bool b) {
    shops.sort((Shop s1, Shop s2) {
      if (!b) {
        // 两项进行交换
        final Shop temp = s1;
        s1 = s2;
        s2 = temp;
      }
      final Comparable<T> s1Value = getField(s1);
      final Comparable<T> s2Value = getField(s2);
      return Comparable.compare(s1Value, s2Value);
    });
    notifyListeners();
  }
}

void onTap() {
  print('data onTap');
}
```

执行以上代码，效果如图 6-12 所示。

图 6-12 通过 PaginatedDataTable 组件实现的带有分页功能的表格

第 7 章

Flutter 常用布局组件

我们知道每个平台应用都有自己的布局方式，例如 Android 应用有线性布局、相对布局、绝对布局、帧布局、表格布局等，Flutter 也不例外。本章将对 Flutter 基础布局组件中的典型布局组件进行讲解，并结合案例带领大家深入学习，内容涉及容器类布局组件、层叠类布局组件、线性布局组件、弹性布局组件和流式布局组件。

7.1 容器类布局组件

容器类布局组件是 Flutter 里非常重要和常用的布局组件。之所以称之为容器类布局，是因为它一般用在外层，负责包裹里面的其他组件，其内部只能有一个子组件。本节将主要介绍 Scaffold 布局组件、Container 布局组件、Center 布局组件。

7.1.1 Scaffold 布局组件

Flutter 布局组件分为两种：只含单个子组件的布局组件，即 SingleChildRenderObjectWidget；具有多个子组件的布局组件，一般有 children 参数，继承自 MultiChildRenderObjectWidget。Flutter 中不同的布局组件对子组件的排列渲染方式不同。

Scaffold 是一个页面布局组件，实现了基本的 Material 风格布局，继承自 StatefulWidget，是有状态组件。我们知道大部分应用页面都是由标题栏、主体内容、底部导航栏、侧滑抽屉菜单等构成，然而每次都重复编写这些内容会大大降低开发效率，所以 Flutter 提供了 Material 风格的 Scaffold 页面布局组件，用于快速搭建页面的基本构成部分。

Scaffold 有以下几个主要属性。

- appBar：显示页面上的标题栏。
- body：构建当前页面的主体内容。
- floatingActionButton：实现页面的主要功能按钮，不配置就不会显示。
- persistentFooterButtons：实现在下方显示的按钮，比如对话框下方的确定按钮、取消按钮。
- drawer：侧滑抽屉菜单控件。
- backgroundColor：设置主体内容的背景颜色。
- bottomNavigationBar：显示页面底部的导航栏。
- resizeToAvoidBottomPadding：避免出现底部“弹出键盘”这类情况遮挡布局。
- bottomSheet：实现底部拉出菜单。

具体可配置的属性参数可通过以下 Scaffold 组件的构造方法和属性参数含义来进一步了解。

```
const Scaffold({
    Key key,
    // 标题栏
    this.appBar,
    // 中间主体内容部分
    this.body,
    // 悬浮按钮
    this.floatingActionButton,
    // 悬浮按钮位置
    this.floatingActionButtonLocation,
    // 悬浮按钮动画
    this.floatingActionButtonAnimator,
    // 在下方显示的按钮
    this.persistentFooterButtons,
    // 侧滑抽屉菜单
    this.drawer,
    this.endDrawer,
    // 底部导航栏
    this.bottomNavigationBar,
    // 底部拉出菜单
    this.bottomSheet,
```

```
    // 设置背景色
    this.backgroundColor,
    this.resizeToAvoidBottomPadding,
    // 重新计算主体内容布局空间大小，避免被遮挡
    this.resizeToAvoidBottomInset,
    // 是否显示在页面顶部，默认为 true
    this.primary = true,
    this.drawerDragStartBehavior = DragStartBehavior.down,
  })
```

有了以上基础，假设我们想在页面中显示底部提示信息组件 Snackbar 或底部滑出风格菜单组件 bottomSheet，可以如下调用。

```
Scaffold.of(context).showSnackBar(new SnackBar(
      content: Text('Hello!'),
    ));

Scaffold.of(context).showBottomSheet...
```

接下来我们来看一个示例：使用 Scaffold 布局组件快速构建一个带有标题栏的页面，代码如下。

```
import 'package:flutter/material.dart';
import 'package:flutter/widgets.dart';

class ScaffoldSamples extends StatefulWidget {
  @override
  State<StatefulWidget> createState() {
    return ScaffoldSamplesState();
  }
}

class ScaffoldSamplesState extends State<ScaffoldSamples> {
  int _selectedIndex = 0;

  @override
  void initState() {
    super.initState();
  }

  @override
```

```
Widget build(BuildContext context) {
  return scaffoldWidget(context);
}

Widget scaffoldWidget(BuildContext context) {
  return Scaffold(
    appBar: AppBar(
      title: Text("标题栏"),
      actions: <Widget>[
        // 导航栏右侧菜单
        IconButton(icon: Icon(Icons.share), onPressed: () {}),
      ],
    ),
    body: Text("Body 内容部分"),
    // 抽屉
    drawer: Drawer(
      child: DrawerHeader(
        child: Text("DrawerHeader"),
      ),
    ),
    // 底部导航栏
    bottomNavigationBar: BottomNavigationBar(
      items: <BottomNavigationBarItem>[
        BottomNavigationBarItem(icon: Icon(Icons.home), title: Text('Home')),
        BottomNavigationBarItem(
            icon: Icon(Icons.category), title: Text('Cagergory')),
        BottomNavigationBarItem(
            icon: Icon(Icons.person), title: Text('Persion')),
      ],
      currentIndex: _selectedIndex,
      fixedColor: Colors.blue,
      onTap: _onItemTap,
    ),
    floatingActionButton: FloatingActionButton(
      child: Icon(Icons.add),
      onPressed: () {
        _onAdd();
      },
    ),
```

```
    );
  }

  void _onItemTap(int index) {
    setState(() {
      _selectedIndex = index;
    });
  }

  void _onAdd() {}
}
```

执行以上代码，效果如图 7-1 所示。

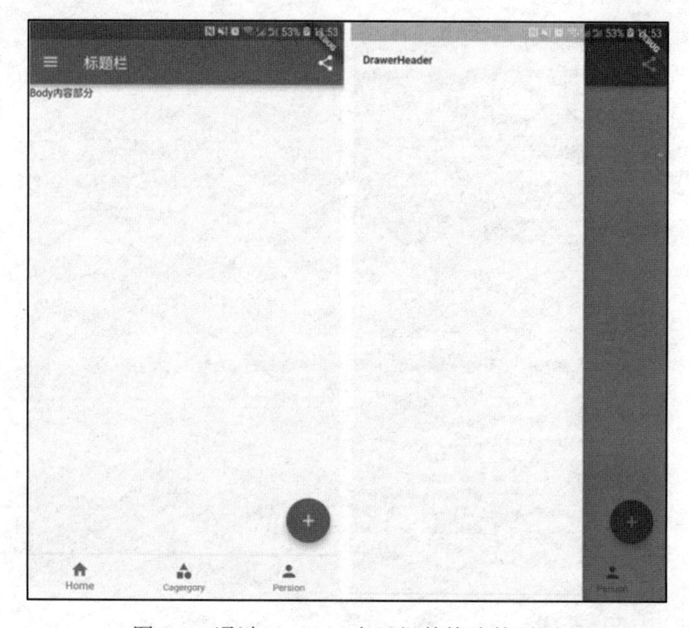

图 7-1　通过 Scaffold 布局组件构建的页面

通过代码可以看出，在 Flutter 中使用 Scaffold 布局组件可以快速构建出一个灵活度非常高的基础页面框架，大家也可以尝试用 Scaffold 组件来配置页面。

7.1.2　Container 布局组件

Container 布局组件可以说是多个小组件的组合容器，可以对 padding、margin、Align、Decoration、Matrix 等属性参数进行设置，用起来很方便，很高效。Container 的构造方法及相关属性参数的含义如下。

```
Container({
    Key key,
    // 子组件对齐方式
    this.alignment,
    // 容器内部 padding
    this.padding,
    // 背景色
    Color color,
    // 背景装饰
    Decoration decoration,
    // 前景装饰
    this.foregroundDecoration,
    // 容器宽度
    double width,
    // 容器高度
    double height,
    // 容器大小的限制条件
    BoxConstraints constraints,
    // 容器外部 margin
    this.margin,
    // 变换，如旋转
    this.transform,
    // 容器内子组件
    this.child,
  })
```

Container 布局组件也非常常用，我们来看一个示例，用 Container 组件实现一个简单的页面，代码如下。

```
body: Container(
        constraints: BoxConstraints.expand(
          height: Theme.of(context).textTheme.display1.fontSize * 1.1 + 200.0,
        ),
        padding: const EdgeInsets.all(8.0),
        // 背景色
        color: Colors.teal.shade700,
        // 子组件居中
        alignment: Alignment.center,
        // 子组件元素
```

```
    child: Text('Hello World',
        style: Theme.of(context)
            .textTheme
            .display1
            .copyWith(color: Colors.white)),
    // 前景装饰
    foregroundDecoration: BoxDecoration(
      image: DecorationImage(
        image: NetworkImage('https://www.example.com/images/frame.png'),
        centerSlice: Rect.fromLTRB(270.0, 180.0, 1360.0, 730.0),
      ),
    ),
    // 旋转
    transform: Matrix4.rotationZ(0.1),
  ),
```

运行以上代码，效果如图 7-2 所示。

图 7-2　Container 布局组件效果

Container 布局组件内部只能放置一个子组件，当然我们也可以在单个子组件中再嵌套组件以达到期望的布局效果。

7.1.3　Center 布局组件

Center 布局组件主要用于实现对齐效果，将内部子组件与自身居中对齐，并根据子组件的大小自动调整自身大小。Center 继承自 Align，Align 继承自 SingleChildRenderObjectWidget，因此 Center 是只含单个子组件的布局组件。

Center 组件的构造方法和属性参数含义如下。

```
Center({
    Key key,
    // 宽度因子
    double widthFactor,
    // 高度因子
    double heightFactor,
    // 子组件
    Widget child
  })
```

大家可能对宽度因子 widthFactor 和高度因子 heightFactor 不太明白，其实它们的作用是，设置 Center 组件的宽度和高度是子组件宽度和高度的多少倍。当然 widthFactor 和 heightFactor 也可以不设置，采用默认设置——宽度横向充满，高度包裹子组件。如果将 widthFactor 和 heightFactor 设置为 2，则表示 Center 组件的宽度和高度是子组件宽度和高度的 2 倍，但是最大不超过屏幕的宽度和高度。

接下来我们通过一个示例来看一下 Center 布局组件的具体用法，代码如下。

```
body: Column(
        children: <Widget>[
          Container(
            color: Colors.blueGrey,
            child: Center(
              widthFactor: 2,
              heightFactor: 2,
              child: Container(
                width: 60,
                height: 30,
                color: Colors.red,
              ),
            ),
          ),
          SizedBox(
            height: 10,
          ),
          Center(
            child: Container(
              width: 60,
```

```
          height: 30,
          color: Colors.teal,
        ),
      ),
      SizedBox(
        height: 10,
      ),
      Center(
        child: Container(
          height: 100.0,
          width: 100.0,
          color: Colors.yellow,
          child: Align(
            // 设置对齐位置约束
            alignment: FractionalOffset(0.2, 0.6),
            child: Container(
              height: 40.0,
              width: 40.0,
              color: Colors.red,
            ),
          ),
        ),
      ),
    ],
  ),
```

执行以上代码，效果如图 7-3 所示。

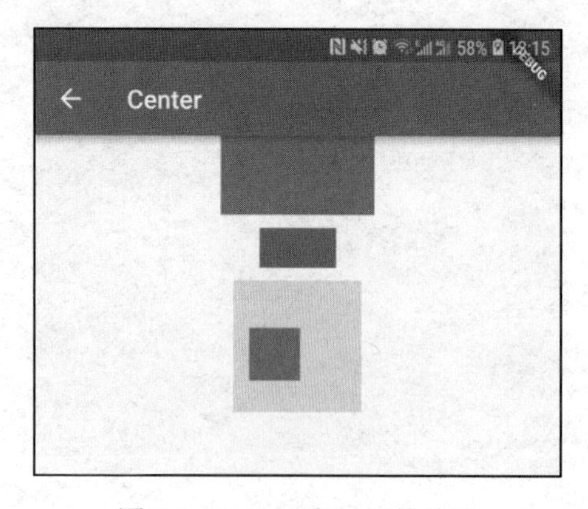

图 7-3　Center 布局组件效果

7.2 层叠类布局组件

层叠类布局组件中的子组件是按照前后顺序堆叠在一起的，类似于堆栈，当然也可以采用绝对定位来进行布局。本节将主要介绍两种具有代表性的层叠类布局组件：Stack 组件和 IndexedStack 组件。

Stack 继承自 MultiChildRenderObjectWidget，内部可以包含多个子组件。而 IndexedStack 继承自 Stack，扩展了 Stack 的一些特性，作用是显示某个特定子组件，令其他子组件不可见。因此 IndexedStack 的尺寸永远与最大的子组件尺寸一致。

Stack 组件的布局行为，取决于子组件是 positioned 还是 non-positioned。

- 对于 positioned 子组件，它们的位置会根据 top、bottom、right 或 left 属性来确定，这几个属性都是相对于 Stack 的左上角来设置的。
- 对于 non-positioned 子组件，它们会根据 alignment 来确定。

另外，Stack 布局中子组件的层级堆叠是有顺序的：最先绘制的子组件在底层，越往后绘制的子组件越靠近顶层。默认情况下，Stack 布局将按照从左上角向外延伸的顺序逐层进行子组件排列。

Stack 的构造方法和属性参数含义如下。

```
Stack({
    Key key,
    // 对齐方式，默认是相对左上角（topStart）设置
    this.alignment = AlignmentDirectional.topStart,
    // 对齐方向
    this.textDirection,
    // 设置无定位子组件尺寸，默认为 loose
    this.fit = StackFit.loose,
    // 超出的部分子组件的处理方式
    this.overflow = Overflow.clip,
    // 子组件
    List<Widget> children = const <Widget>[],
  })
```

Stack 的属性不多，首先看一下 alignment 属性，alignment 属性主要用来配置 Stack 内部子组件的对齐方式，默认以左上角为基准，其构造方法和属性参数含义如下。

```
// 王轴顶部对齐，交叉轴偏左
static const Alignment topLeft = Alignment(-1.0, -1.0);
// 主轴顶部对齐，交叉轴居中
static const Alignment topCenter = Alignment(0.0, -1.0);
// 主轴顶部对齐，交叉轴偏右
static const Alignment topRight = Alignment(1.0, -1.0);
// 主轴居中，交叉轴偏左
static const Alignment centerLeft = Alignment(-1.0, 0.0);
// 居中
static const Alignment center = Alignment(0.0, 0.0);
// 主轴居中，交叉轴偏右
static const Alignment centerRight = Alignment(1.0, 0.0);
// 主轴底部对齐，交叉轴偏左
static const Alignment bottomLeft = Alignment(-1.0, 1.0);
// 主轴底部对齐，交叉轴居中
static const Alignment bottomCenter = Alignment(0.0, 1.0);
// 主轴底部对齐，交叉轴偏右
static const Alignment bottomRight = Alignment(1.0, 1.0);
```

接下来再看一下 fit 属性。fit 属性主要用来配置子组件的填充模式，其构造方法和属性参数含义如下。

```
// 子组件宽松的取值，可以从 min 到 max
loose,
// 子组件尽可能地占用剩余空间，取 max 尺寸
expand,
// 不改变子组件的约束条件
passthrough,
```

最后我们看一下 overflow 属性，overflow 属性主要用来配置子组件超出容器部分的处理方式，其构造方法和属性参数含义如下。

```
// 超出部分不会被裁剪，正常显示
visible,
// 超出部分会被裁剪
clip,
```

熟悉了 Stack 组件的基本用法和属性参数后，再学习 IndexedStack 布局组件就容易多了。IndexedStack 组件是 Stack 布局组件的继承拓展，可以进行层叠子组件前后顺序的配置，其构造方法和属性参数含义如下。

```
IndexedStack({
```

```
    Key key,
    AlignmentGeometry alignment = AlignmentDirectional.topStart,
    TextDirection textDirection,
    StackFit sizing = StackFit.loose,
    // 有索引的组件显示，其他组件隐藏
    this.index = 0,
    // 子组件
    List<Widget> children = const <Widget>[],
  })
```

可以看到，IndexedStack 组件的属性参数和 Stack 组件基本一致，很容易理解。接下来我们通过一个示例来理解 Stack 组件和 IndexedStack 组件的具体用法：结合两者实现一个布局组件堆叠效果，代码如下。

```
body: Column(
          children: <Widget>[
            // Stack 层叠布局
            Stack(
              children: <Widget>[
                Container(
                  width: 300,
                  height: 300,
                  color: Colors.grey,
                ),
                Container(
                  width: 200,
                  height: 200,
                  color: Colors.teal,
                ),
                Container(
                  width: 100,
                  height: 100,
                  color: Colors.blue,
                ),
                Text(
                  "Stack",
                  style: TextStyle(color: Colors.white),
                ),
              ],
            ),
```

```
        SizedBox(
          height: 10,
        ),
        // IndexedStack 层叠布局
        IndexedStack(
          // 指定显示的子组件序号，其余子组件隐藏
          index: 2,
          children: <Widget>[
            Container(
              width: 300,
              height: 300,
              color: Colors.grey,
            ),
            Container(
              width: 200,
              height: 200,
              color: Colors.teal,
            ),
            Container(
              width: 100,
              height: 100,
              color: Colors.blue,
            ),
            Text(
              "Stack",
              style: TextStyle(color: Colors.white),
            ),
          ],
        )
      ],
    )
```

执行以上代码，效果如图 7-4 所示。

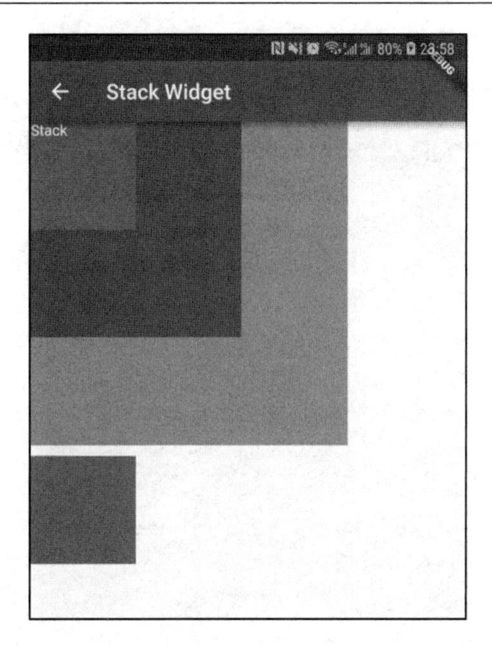

图 7-4　Stack 组件结合 IndexedStack 组件效果

我们可以看出，Stack 组件主要用来进行堆叠式帧布局，而 IndexedStack 组件只能用来设置堆叠的帧布局中子组件的层级顺序。

7.3　线性布局组件

Flutter 中的线性布局组件主要用于在垂直方向或水平方向排布子组件，其中比较重要的是 Row 布局组件和 Column 布局组件，本节我们将主要介绍这两种组件。

7.3.1　Row 布局组件

Row 布局组件类似于 Android 中的 LinearLayout 线性布局，用来实现水平布局，里面的 children 子组件按照水平方向排列。Row 的继承关系如下。

```
Row -> Flex -> MultiChildRenderObjectWidget -> RenderObjectWidget ...
```

可以看出 Row 是 Flex 的拓展子类，它的内部可以包含多个子组件。Row 的构造方法和属性参数含义如下。

```
Row({
    Key key,
```

```
    // 主轴（Row 的主轴是横向轴）方向上的对齐方式
    MainAxisAlignment mainAxisAlignment = MainAxisAlignment.start,
    // 在主轴方向占有空间的值，默认是 max
    MainAxisSize mainAxisSize = MainAxisSize.max,
    // 交叉轴方向上的对齐方式，Row 的高度等于子组件中最高的子组件的高度
    CrossAxisAlignment crossAxisAlignment = CrossAxisAlignment.center,
    // 水平方向子组件的排列方向：从左到右排列还是反向
    TextDirection textDirection,
   // 垂直对齐排列方向，默认是 VerticalDirection.down，表示从上到下
   // 这个参数一般用于 Column 组件里
    VerticalDirection verticalDirection = VerticalDirection.down,
    // 字符对齐基线方式
    TextBaseline textBaseline,
    // 子组件集合
    List<Widget> children = const <Widget>[],
  })
```

Row 布局组件的构造方法不算多，我们通过配置这些属性就可以实现一些常见的水平排列布局效果。下面我们着重看一下 Row 的主轴和交叉轴属性。

MainAxisAlignment 是 Row 的主轴属性，通过这个属性我们可以配置子组件在主轴方向上的对齐方式，具体可以配置的值如下。

```
enum MainAxisAlignment {
  // 与主轴起点对齐
  start,
  // 将子组件放置在主轴的末尾，按照末尾对齐
  end,
  // 将子组件放置在主轴中心对齐
  center,
  // 将主轴方向上的空白区域均分，使得子组件之间的空白区域相等
 // 首尾子组件都靠近首尾，没有间隙，类似于两端对齐
  spaceBetween,
  // 将主轴方向上的空白区域均分，使得子组件之间的空白区域相等，但是首尾子组件的空白区域为 1/2
  spaceAround,
  // 将主轴方向上的空白区域均分，使得子组件之间的空白区域均相等
  spaceEvenly,
}
```

再来看一下 Row 布局组件的交叉轴属性。CrossAxisAlignment 是 Row 的交叉轴属性，用来设置子组件在交叉轴方向的对齐方式。Row 组件的高度等于最高的子组件的高度。

CrossAxisAlignment 属性参数可以配置的值如下。

```
enum CrossAxisAlignment {
  // 将子组件在交叉轴上起点处展示
  start,
  // 将子组件在交叉轴上末尾处展示
  end,
  // 将子组件在交叉轴上居中展示
  center,
  // 让子组件填满交叉轴方向
  stretch,
  // 在交叉轴方向，使得子组件与 baseline（基准线）对齐
  baseline,
}
```

与以上两个属性类似的还有 MainAxisSize 属性，MainAxisSize 属性用来配置 Row 主轴方向上子组件占有空间的方式。Row 的主轴是水平轴，主轴方向上子组件占有空间的默认方式是 max，即按照最大化方式占用主轴方向上的可用空间。MainAxisSize 属性参数可以配置的值如下。

```
enum MainAxisSize {
  // 根据传入的布局约束条件，最大化占用主轴方向的可用空间，尽可能充满可用宽度
  max,
  // 与 max 相反，最小化占用主轴方向的可用空间
  min,
}
```

以上就是 Row 组件的几个核心属性。有了这些做支撑，我们就可以进行 Row 组件实践了。接下来我们通过一个示例来深入理解 Row 组件的布局特点，代码如下。

```
Column(
        children: <Widget>[
          // 默认横向排列元素
          Row(
            verticalDirection: VerticalDirection.up,
            textBaseline: TextBaseline.ideographic,
            children: <Widget>[
              RaisedButton(
                color: Colors.blue,
                child: Text(
                  '我是按钮一\n 按钮',
```

```
                style: TextStyle(color: Colors.white),
              ),
              onPressed: () {},
            ),
            RaisedButton(
              color: Colors.grey,
              child: Text(
                '   我是按钮二   ',
                style: TextStyle(color: Colors.black),
              ),
              onPressed: () {},
            ),
            RaisedButton(
              color: Colors.orange,
              child: Text(
                '      我是按钮三      ',
                style: TextStyle(color: Colors.white),
              ),
              onPressed: () {},
            ),
          ],
        ),
        SizedBox(
          height: 10,
        ),
        // 默认横向排列元素
        Row(
          children: <Widget>[
            const FlutterLogo(),
            const Expanded(
              child: Text(
                  'Flutter\'s hot reload helps you quickly and easily experiment, b
uild UIs, add features, and fix bug faster. Experience sub-second reload times, witho
ut losing state, on emulators, simulators, and hardware for iOS and Android.'),
            ),
            const Icon(Icons.sentiment_very_satisfied),
          ],
        ),
        SizedBox(
          height: 10,
```

```
          ),
          // 居中排列元素
          Row(
            mainAxisAlignment: MainAxisAlignment.center,
            children: <Widget>[
              Text(
                " 我们居中显示 |",
                style: TextStyle(color: Colors.teal),
              ),
              Text(" Flutter 的 Row 布局组件 "),
            ],
          ),
        ],
      ),
```

执行以上代码，效果如图 7-5 所示。可以看到，Row 布局组件一般用来实现子组件水平排列的布局效果，大家如果遇到这种情景，可以考虑使用 Row 来实现。

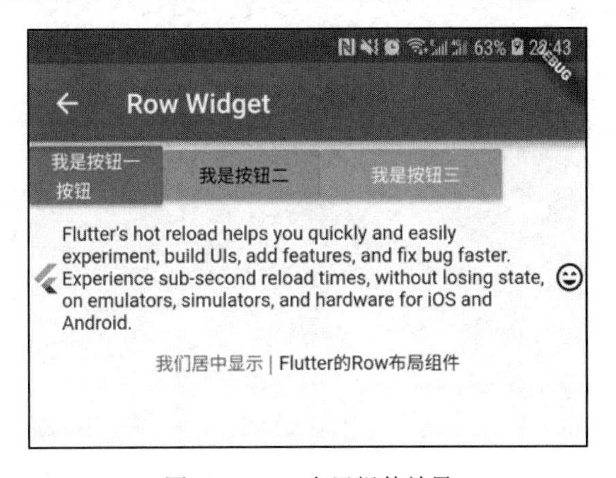

图 7-5 Row 布局组件效果

7.3.2 Column 布局组件

有了 Row 布局组件的基础，再学习 Column 布局组件将很容易。与 Row 相对，Column 主要用于实现垂直排列子组件的布局效果，其继承关系如下。

```
Column -> Flex -> MultiChildRenderObjectWidget -> RenderObjectWidget ...
```

Column 组件也是 Flex 组件的拓展子类，它的内部可以包含多个子组件，子组件是垂直排列的。

在 7.3.1 节中，我们详细介绍了 Row 的构造方法和属性参数含义，Column 的构造方法和属性参数含义与 Row 是一致的，但其主轴和交叉轴的属性与 Row 相反，大家可以参考 7.3.1 节的内容自行理解，这里不再重复讲解。我们直接来看一个 Column 布局的示例，代码如下。

```
Column(
        // 纵向排列子组件
        children: <Widget>[
          RaisedButton(
            color: Colors.blue,
            child: Text(
              '我是按钮一',
              style: TextStyle(color: Colors.white),
            ),
            onPressed: () {},
          ),
          RaisedButton(
            color: Colors.grey,
            child: Text(
              '    我是按钮二    ',
              style: TextStyle(color: Colors.black),
            ),
            onPressed: () {},
          ),
          RaisedButton(
            color: Colors.orange,
            child: Text(
              '       我是按钮三       ',
              style: TextStyle(color: Colors.white),
            ),
            onPressed: () {},
          ),
          SizedBox(
            height: 10,
          ),
          const FlutterLogo(),
          Text(
              'Flutter\'s hot reload helps you quickly and easily experiment, build U
Is, add features, and fix bug faster. Experience sub-second reload times, without los
ing state, on emulators, simulators, and hardware for iOS and Android.'),
          const Icon(Icons.sentiment_very_satisfied),
```

```
        ],
      ),
```

执行以上代码，效果如图 7-6 所示。

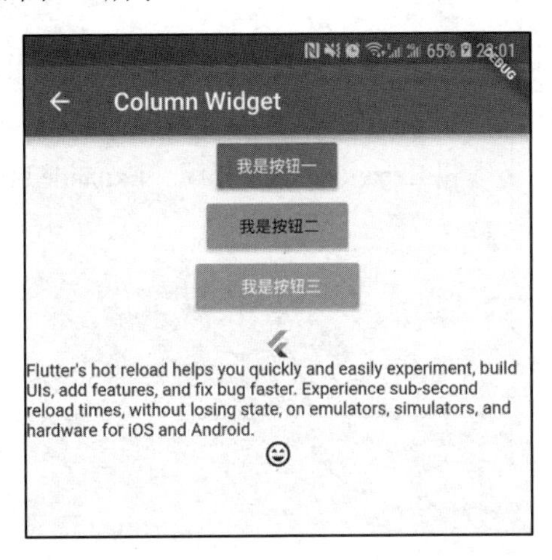

图 7-6 Column 布局组件效果

7.4 弹性布局组件

Flutter 的弹性布局是指，按照权重比例弹性设置动态宽度和高度，其中 Flex 和 Expanded 是弹性布局中最典型的布局组件。

Flex 组件是 Row 组件和 Column 组件的父类，可以按照一定的比例分配布局空间。Flex 组件继承自 MultiChildRenderObjectWidget，内部可以包含多个子组件，子组件按照配置的权重比例布局排列。Flex 组件一般和 Expanded 组件搭配使用，Expanded 组件主要用于让子组件扩展占用 Flex 的剩余空间。

接下来，我们看一下 Flex 组件的构造方法和属性参数含义，如下。

```
Flex({
    Key key,
    // 子组件排列方向：横向或纵向
    @required this.direction,
    this.mainAxisAlignment = MainAxisAlignment.start,
    this.mainAxisSize = MainAxisSize.max,
    this.crossAxisAlignment = CrossAxisAlignment.center,
    this.textDirection,
```

```
    this.verticalDirection = VerticalDirection.down,
    this.textBaseline,
    List<Widget> children = const <Widget>[],
  })
```

Flex 组件的属性参数和 Column 组件大致相同，这里不再重复讲解。其实单独学习 Flex 组件没有意义，因为我们一般会直接用它的子类 Row 组件和 Column 组件来进行布局。如果想要使用 Flex 组件的权重特性，必须和 Expanded 组件搭配。Expanded 组件的构造方法和属性参数含义如下。

```
const Expanded({
    Key key,
    // 占用空间比重、权重
    int flex = 1,
    // 子组件
    @required Widget child,
  })
```

Expanded 组件的核心就是设置 flex，即占用空间比重、权重，其内部需要包裹一个子组件，设置的属性对这个子组件生效。

下面我们通过一个示例看一下 Flex 组件和 Expanded 组件的搭配用法，代码如下。

```
body: Row(
          children: <Widget>[
            Expanded(
              // 设置 flex，这里是占用 2/5 空间
              flex: 2,
              child: RaisedButton(
                color: Colors.blue,
                child: Text(
                  '我是按钮一',
                  style: TextStyle(color: Colors.white),
                ),
                onPressed: () {},
              ),
            ),
            // 设置 flex，这里是占 1/5 空间
            Expanded(
              flex: 1,
              child: Column(
```

```
          children: <Widget>[
            Expanded(
              flex: 2,
              child: RaisedButton(
                color: Colors.grey,
                child: Text(
                  '    我是按钮一    ',
                  style: TextStyle(color: Colors.black),
                ),
                onPressed: () {},
              ),
            ),
            Expanded(
              flex: 1,
              child: RaisedButton(
                color: Colors.teal,
                child: Text(
                  '    我是按钮二    ',
                  style: TextStyle(color: Colors.white),
                ),
                onPressed: () {},
              ),
            )
          ],
        ),
      ),
      // 设置 flex，这里是占 2/5 空间
      Expanded(
        flex: 2,
        child: RaisedButton(
          color: Colors.grey,
          child: Text(
            '    我是按钮二    ',
            style: TextStyle(color: Colors.black),
          ),
          onPressed: () {},
        ),
      )
    ],
```

```
    )
```

执行以上代码，效果图如图 7-7 所示。

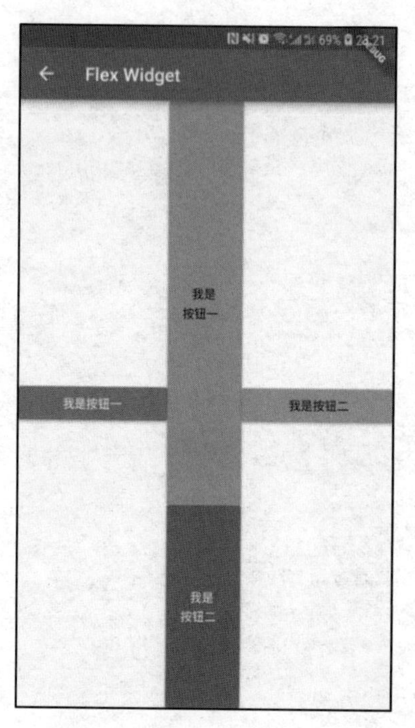

图 7-7　Flex 组件搭配 Expanded 组件布局效果

在实际开发时，首选 Column 和 Row 线性布局组件，如果不能满足需求，可以考虑使用 Flex 组件搭配 Expanded 组件来实现期望的效果。

7.5　流式布局组件

本节将介绍流式布局组件，也叫瀑布布局组件。在 Flutter 中，我们主要通过 Flow 布局组件和 Wrap 布局组件来实现流式布局。流式布局的主要作用是，当某个组件因不能自动换行而溢出屏幕或被裁剪时，及时进行自动换行，有点类似于文字的自动换行。

7.5.1　Flow 布局组件

Flow 布局组件可以自定义规则以控制子组件的布局排列。Flow 性能非常好，继承自 MultiChildRenderObjectWidget，实现布局也比较灵活，可以实现定制的布局效果。

Flow 组件的构造方法和属性参数含义如下。

```
Flow({
    Key key,
    // 子组件布局排列规则
    @required this.delegate,
    // 子组件
    List<Widget> children = const <Widget>[],
  })
```

Flow 的构造方法很简单，最重要的就是 delegate 配置规则。通过 delegate 配置规则，我们可以自己实现布局内子组件的排列方式。使用时，我们需要实现一个 Delegate 组件，继承自 FlowDelegate 组件，然后编写具体的布局排列代码。

以下是一个使用 Flow 布局组件的完整示例：实现一个自定义排列方式的流式布局。

```
class FlowSamplesState extends State<FlowSamples> {
  @override
  void initState() {
    super.initState();
  }

  @override
  Widget build(BuildContext context) {
    return Scaffold(
        appBar: AppBar(title: Text('Flow Demo'), primary: true),
        body: Flow(
          // 子组件布局排列规则
          delegate: FlowWidgetDelegate(margin: EdgeInsets.all(10.0)),
          children: <Widget>[
            Container(
              width: 80.0,
              height: 80.0,
              color: Colors.orange,
            ),
            Container(
              width: 160.0,
              height: 80.0,
              color: Colors.teal,
            ),
            Container(
```

```
            width: 80.0,
            height: 80.0,
            color: Colors.red,
          ),
          Container(
            width: 80.0,
            height: 80.0,
            color: Colors.yellow,
          ),
          Container(
            width: 80.0,
            height: 80.0,
            color: Colors.brown,
          ),
          Container(
            width: 80.0,
            height: 80.0,
            color: Colors.purple,
          ),
        ],
      ));
  }
}

class FlowWidgetDelegate extends FlowDelegate {
  EdgeInsets margin = EdgeInsets.all(10);
  // 构造方法，传入每个 child 的间隔
  FlowWidgetDelegate({this.margin});

  // 必须要重写的方法
  @override
  void paintChildren(FlowPaintingContext context) {
    var screenWidth = context.size.width;
    double offsetX = margin.left; // 记录横向绘制的 x 轴坐标
    double offsetY = margin.top; // 记录纵向绘制的 y 轴坐标
    // 遍历子组件
    for (int i = 0; i < context.childCount; i++) {
      // 如果当前 x 左边加上子组件宽度小于屏幕宽度则继续绘制，否则换行
      var width = context.getChildSize(i).width + offsetX + margin.right;
```

```
      if (width < screenWidth) {
        // 绘制子组件
        context.paintChild(i,
            transform: Matrix4.translationValues(offsetX, offsetY, 0.0));
        offsetX = width + margin.left;
      } else {
        offsetX = margin.left;
        offsetY += context.getChildSize(i).height + margin.top + margin.bottom;
        // 绘制子组件
        context.paintChild(i,
            transform: Matrix4.translationValues(offsetX, offsetY, 0.0));
        offsetX += context.getChildSize(i).width + margin.left + margin.right;
      }
    }
  }

  // 必须要重写的方法：是否需要重绘
  @override
  bool shouldRepaint(FlowDelegate oldDelegate) {
    return oldDelegate != this;
  }

  // 可选重写方法：是否需要重新布局
  @override
  bool shouldRelayout(FlowDelegate oldDelegate) {
    return super.shouldRelayout(oldDelegate);
  }

  // 可选重写方法：设置 Flow 布局组件的尺寸
  @override
  Size getSize(BoxConstraints constraints) {
    return super.getSize(constraints);
  }

  // 可选重写方法：设置每个 child 的布局约束条件
  @override
  BoxConstraints getConstraintsForChild(int i, BoxConstraints constraints) {
    return super.getConstraintsForChild(i, constraints);
  }
```

```
}
```

执行以上代码，效果如图 7-8 所示。

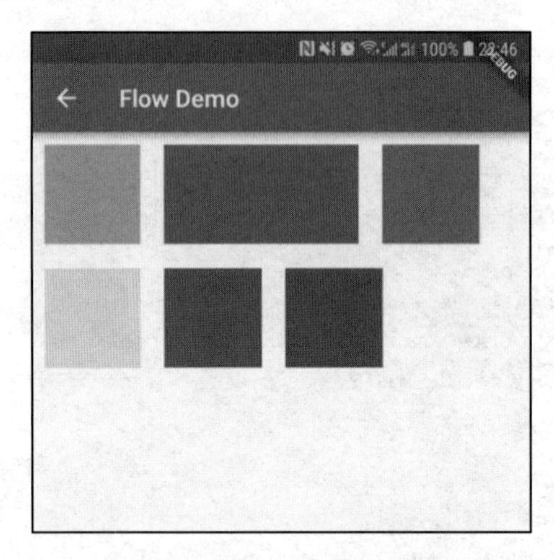

图 7-8　Flow 布局组件效果

通过以上示例可以看到，使用 Flow 布局组件的重点就是要重写一个 FlowDelegate 组件代码，FlowDelegate 里面最重要的方法是 paintChildren 方法，用于定制子组件的排列方式。当然我们也可以实现更复杂一点的效果，可以根据自己的需求排列子组件。

7.5.2　Wrap 布局组件

在 Flutter 中，Wrap 布局组件的功能是使子组件自动换行，默认情况下内部子组件的排列方向是水平的，当超过指定宽度时就自动换到下一行。例如我们平时看到的商品标签页，那些标签是自动按行排列的。

其实 Wrap 组件实现的效果，Flow 组件也可以实现。我们可以根据实际情况合理选择。Wrap 也继承自 MultiChildRenderObjectWidget，即内部可以放置多个子组件。Wrap 组件的构造方法和属性参数含义如下。

```
Wrap({
    Key key,
    // 子组件在主轴上的排列方向
    this.direction = Axis.horizontal,
    // 子组件在主轴方向上的对齐方式
    this.alignment = WrapAlignment.start,
    // 子组件在主轴方向上的间距
```

```
    this.spacing = 0.0,
    // 子组件在纵轴方向上的对齐方式
    this.runAlignment = WrapAlignment.start,
    // 子组件在纵轴方向上的间距
    this.runSpacing = 0.0,
    // 子组件在交叉轴上的对齐方式
    this.crossAxisAlignment = WrapCrossAlignment.start,
    // 文本方向
    this.textDirection,
    // 排列方式为 Vertical 时，子组件的排列顺序
    this.verticalDirection = VerticalDirection.down,
    // 子组件集合
    List<Widget> children = const <Widget>[],
  })
```

属性参数虽然多，但是很容易理解。Wrap 的很多属性参数和 Row、Column、Flex 等组件有重复的地方，大家可以对比着理解。接下来通过一个示例来看一下 Wrap 的特点和用法，代码如下。

```
class WrapSamplesState extends State<WrapSamples> {
  @override
  void initState() {
    super.initState();
  }

  @override
  Widget build(BuildContext context) {
    return Scaffold(
        appBar: AppBar(title: Text('Wrap Demo'), primary: true),
        // Wrap 包裹的子组件会自动换行
        body: Wrap(
          spacing: 8.0, // 子组件横向间距
          runSpacing: 4.0, // 子组件纵向间距（每行间距）
          // 子组件元素集合
          children: <Widget>[
            Chip(
              avatar: CircleAvatar(
                  backgroundColor: Colors.blue.shade900, child: Text('A')),
              label: Text('全部'),
            ),
```

```
          … …
          Chip(
            avatar: CircleAvatar(
                backgroundColor: Colors.blue.shade900, child: Text('F')),
            label: Text('认证作者 23'),
          ),
          Chip(
            avatar: CircleAvatar(
                backgroundColor: Colors.blue.shade900, child: Text('G')),
            label: Text('同城 1496'),
          ),
        ],
      ));
  }
}
```

执行以上代码，效果如图 7-9 所示。

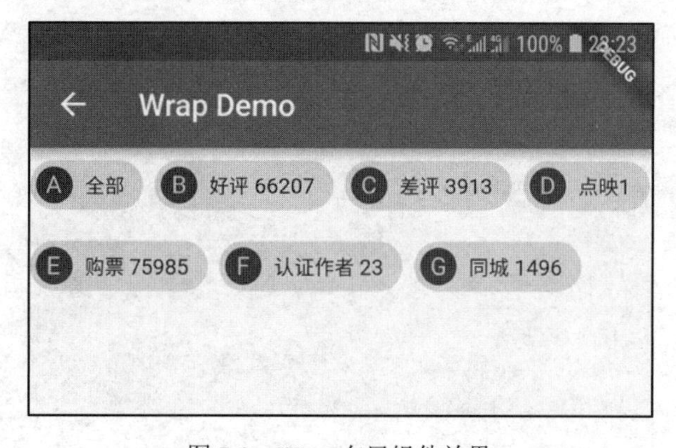

图 7-9　Wrap 布局组件效果

我们通过 Wrap 组件可以轻松实现类似于美团商品标签评论页的效果。当流式布局效果不是非常复杂时，Wrap 组件用起来要简单些。如果用 Wrap 组件实现不了，那么就可以考虑用 Flow 组件来实现。大家可以多多实践，巩固流式布局组件的相关用法。

第 8 章

Flutter 路由与生命周期

在前面几章中，我们着重介绍了 Flutter 中的常用组件，从本章开始，我们将讲解 Flutter 中一些常用的技术要点。

本章将主要讲解 Flutter 中路由的概念、使用方法，以及 Flutter 的生命周期。在 Flutter 中，路由负责页面跳转和数据传递，而生命周期主要是指一个页面从创建到销毁的整个状态管理过程。

8.1 路由简介

在 Flutter 中，路由主要用来处理页面跳转、数据传递等操作，常和 Navigator 配合使用。Navigator 主要负责路由页面的堆栈管理和操作，如添加跳转页面、移除页面等。

在 Flutter 中，路由跳转的实现方式主要有两种：一种是在页面的 MaterialApp 里配置提前定义好的路由列表，称为静态路由；另一种就是在要用到跳转的地方编写代码并传递参数，称为动态路由。

Flutter 中的静态路由实现方式是，提前把各个需要跳转的页面路径注册在路由表中。静态路由不支持向目标页面传递参数，但是可以接收目标页面的返回值。

动态路由相对来说比较灵活，它可以在需要进行页面跳转的地方使用，并且支持向目标页面传递参数及参数回传，在使用时也不需要提前规划好页面路径，只需要在具体跳转逻辑中构造 MaterialPageRoute 对象或者 PageRouterBuilder 对象即可，还支持设置页面跳转动画。

Flutter 路由的核心是 Navigator，Navigator 负责管理路由堆栈并提供管理路由堆栈的方法，如 Navigator.push 和 Navigator.pop 方法。我们可以通过 Navigator 来管理路由对象的进栈、出栈，从一个页面跳转到另一个页面。

Navigator 继承自 StatefulWidget，属于有状态组件，其构造方法很简单，核心就是明确路由跳转规则，具体如下。

```
const Navigator({
    Key key,
    // 默认路由
    this.initialRoute,
    // 通用路由跳转规则
    @required this.onGenerateRoute,
    // 未知路由
    this.onUnknownRoute,
    // 路由监听
    this.observers = const <NavigatorObserver>[]
  })
```

我们再看一下 Navigator 中的主要方法，具体如下。

- push：自定义使用 Route 子类（如 MaterialPageRoute 子类）实现个性化页面跳转。
- of：实例化 Navigator。
- pop：从路由堆栈里移除页面组件，退出页面。
- canPop：判断当前页面能否进行 pop 操作，并返回 bool 值。
- maybePop：判断能否弹出，当首页仅剩一个页面时不进行弹出操作。
- popAndPushNamed：将当前页面从路由堆栈移除，然后跳转到指定名称页面。
- popUntil：返回跳转，直到指定页面便停止。
- pushAndRemoveUntil：移除所有页面，然后跳转到指定页面。
- pushNamed：通过命名好的路由实现页面跳转。
- pushNamedAndRemoveUntil：将指定名称的页面加入路由，将之前路由堆栈里的页面访问路径逐个移除，直到访问到指定名称的页面为止。

- pushReplacement：将路由堆栈中的当前页面位置替换成跳转页面。
- pushReplacementNamed：将路由堆栈中当前页面的位置替换成已命名的跳转页面。
- removeRoute：移除路由。
- replace：替换路由。

我们可以通过调用 Navigator 中的方法来实现 Flutter 中的路由导航功能，非常方便和简单。

8.2　路由跳转

上一节我们简单介绍了 Flutter 路由的概念，本节我们一起来看一下 Flutter 中的路由跳转是如何实现的。

通过静态路由实现页面跳转时，要在 MaterialApp 的 routes 参数里配置定义好的路由列表，构造方法如下，其中第一个参数填写我们定义的目标页面别名，第二个参数填写要跳转的目标页面的组件实例。

```
Map<String, WidgetBuilder> routes;
```

我们需要先在页面的 MaterialApp 里提前定义好路由跳转路径，配置代码如下。

```
return MaterialApp(
      title: 'Flutter Demo',
      theme: ThemeData(
        primarySwatch: Colors.teal,
      ),
      home: ShowAppPage(),
      // 配置路由列表
      routes: <String, WidgetBuilder>{
        // 第一个参数填写目标页面别名，第二个参数为目标页面组件实例
        '/buttonpage': (BuildContext context) {
          return ButtonSamples();
        },
        '/routepage': (BuildContext context) {
          return RouteSamples();
        },
      },
    );
```

```
// 可以用 lambda 简写路由
routes: <String, WidgetBuilder>{
        '/buttonpage': (BuildContext context) => ButtonSamples(),
        '/routepage': (BuildContext context) => RouteSamples(),
      },
```

接下来就可以通过设置好的目标页面别名来执行跳转了，这里要用到 Navigator，代码如下。

```
...
body: Center(
        child: FlatButton(
          child: Text("路由跳转"),
          onPressed: () {
            // 通过之前定义的目标页面别名来执行跳转
            Navigator.of(context).pushNamed("buttonpage");
          },
        ),
      )
```

可以看到，这种静态路由配置方式需要在两个地方进行处理，一个是配置路由列表，另一个是在跳转的地方编写逻辑，相对来说麻烦一些。

接下来我们介绍更灵活的第二种路由使用方法：在需要跳转的地方直接配置路由，即动态路由。这里我们要结合 Navigator 与 PageRoute 来实现，并且传递参数，简单示例如下。

```
Navigator.push(context,MaterialPageRoute(builder: (context) {
return SearchPage();
}
));
```

这种方法实际上就是用 Navigator 结合 PageRoute 来实现路由跳转。PageRoute 可以用两个实例对象来构建：使用 MaterialPageRoute 来构建；使用 PageRouteBuilder 来构建。

我们先来看一下 MaterialPageRoute 的构造方法及属性参数含义，如下。

```
MaterialPageRoute({
    // 构建页面
    @required this.builder,
    // 路由设置
    RouteSettings settings,
    // 是否保存页面状态、内容
    this.maintainState = true,
```

```
  bool fullscreenDialog = false,
})
```

MaterialPageRoute 构造方法的核心是 builder，即构建路由跳转的页面。再来看一下 PageRouteBuilder 构造方法及属性参数含义，如下。

```
PageRouteBuilder({
    // 路由设置
    RouteSettings settings,
    // 目标页面
    @required this.pageBuilder,
    // 跳转动画设置
    this.transitionsBuilder = _defaultTransitionsBuilder,
    this.transitionDuration = const Duration(milliseconds: 300),
    this.opaque = true,
    this.barrierDismissible = false,
    this.barrierColor,
    this.barrierLabel,
    this.maintainState = true,
})
```

使用 PageRouteBuilder 构建的路由可以设置跳转动画。

了解了构造方法，我们来看具体的示例。使用 MaterialPageRoute 时无须预先配置路由目标页面和名称，代码如下。

```
FlatButton(
      child: Text("路由跳转"),
      onPressed: () {
        Navigator.of(context)
            .push(MaterialPageRoute(builder: (BuildContext context) {
          return ButtonSamples();
        }));
      },
    );
```

通过 PageRouteBuilder 可以创建一个更加丰富、复杂的页面路由，设置跳转动画代码如下。

```
Navigator.push(context, PageRouteBuilder(
  opaque: false,
  pageBuilder: (BuildContext context, Animation<double> animation,
             Animation<double> secondaryAnimation) {
    return ButtonSamples();
  },
  transitionsBuilder: (BuildContext context, Animation<double> animation, Animation<
double> secondaryAnimation, Widget child) {
    return FadeTransition(
      opacity: animation,
      child: RotationTransition(
        turns: Tween<double>(begin: 0.5, end: 1.0).animate(animation),
        child: child,// 这里的 child 是 pageBuilder 里返回的目标页面
      ),
    );
  }
));
```

通过上面的代码我们就可以在路由跳转时添加一个自定义的跳转动画，提升用户体验。以上代码的核心是配置 transitionsBuilder，这个参数是配置自定义跳转动画的。

最后再看一个复杂一些的路由跳转情况。例如我们有一个首页 Tab 页面，内部有自己的路由跳转信息，当我们跳转到个人中心时，需要判断是否已登录，没有登录就跳转到登录页，登录过就直接显示个人中心信息。这是一个路由里面嵌套路由的典型例子，实现方法如下。

```
class MyApp extends StatelessWidget {
  @override
  Widget build(BuildContext context) {
    return MaterialApp(
      initialRoute: '/',
      routes: {
        '/': (BuildContext context) => HomePage(),
        // 登录页
        '/signup': (BuildContext context) => SignUpPage(),
      },
    );
  }
}
```

```
class SignUpPage extends StatelessWidget {
 @override
 Widget build(BuildContext context) {
   // SignUpPage 有内部路由
   return Navigator(
     // 默认路由页面
     initialRoute: 'signup/personal_info',
     // 内部路由的跳转处理
     onGenerateRoute: (RouteSettings settings) {
       WidgetBuilder builder;
       switch (settings.name) {
         // 跳转到个人中心
         case 'signup/personal_info':
           builder = (BuildContext _) => CollectPersonalInfoPage();
           break;
         // 跳转到登录页
         case 'signup/choose_credentials':
           builder = (BuildContext _) => ChooseCredentialsPage();
           break;
         default:
           throw Exception('Invalid route: ${settings.name}');
       }
       return MaterialPageRoute(builder: builder, settings: settings);
     },
   );
 }
}
```

以上我们已经对 Flutter 路由跳转的几种实现方式和不同情况都进行了分析和讲解，大家可以仔细阅读代码并进行实践，巩固 Flutter 路由相关知识点。

8.3　参数传递

本节我们介绍 Flutter 中页面跳转的同时传递参数的实现方法。

对于页面跳转和参数传递，这里建议使用动态路由，而不使用在 MaterialApp 的 Routes 属性里进行静态定义的方式。为了对比两种方式，我们先来看一下静态定义的参数传递方法，具体如下。

```
// 通过 key/value 传递参数
Navigator.pushNamed(
      context,
      '/weather',
      arguments: <String, String>{
        'city': 'Berlin',
        'country': 'Germany',
      },
    );
// 通过 Object 传递参数
class WeatherRouteArguments {
  WeatherRouteArguments({this.city, this.country});
  final String city;
  final String country;
}

Navigator.pushNamed(
    context,
    '/weather',
    arguments: WeatherRouteArguments(city: 'Berlin', country: 'Germany'),
  );

// 通过 RouteSettings 接收页面传递的参数，不灵活
MaterialApp(
        onGenerateRoute: (RouteSettings settings){
            ...
            settings.arguments
            ...
},
```

通过以上代码，我们可以看出，静态定义的参数传递方式比较麻烦，要编写的内容比较多，不够灵活，容易出错。因此，通常我们通过动态路由的方式进行参数传递，示例如下。

```
Navigator.push(context, new MaterialPageRoute(builder: (BuildContext context){
  return new ButtonSamples(
    title:'标题',
    name:'名称'
  );
}))
```

```
...
// 在第二个页面构造方法里接收参数
class ButtonSamples extends StatefulWidget {
  String title;
  String name;
  ButtonSamples({Key key, this.title, this.name}) : super(key: key);

  @override
  State<StatefulWidget> createState() {
    return ButtonSamplesState();
  }
}
```

或者，我们也可以传递一个 Object 对象，代码如下。

```
class Book{
  String title;
  String name;

  Book({this.title, this.name});
}

Navigator.push(context, new MaterialPageRoute(builder: (BuildContext context){
  return new ButtonSamples(Book
    title:'标题',
    name:'名称'
  );
}))

// 在第二个页面构造方法里接收参数
class ButtonSamples extends StatefulWidget {
  Book book;

  ButtonSamples({Key key, this.book}) : super(key: key);

  @override
  State<StatefulWidget> createState() {
    return ButtonSamplesState();
  }
}
```

可以看出，动态路由传递参数只需要在目标页面配置要传递的参数，而在起始页面只要将所需的参数传递进来即可。这种方式配置简单、清晰、方便，使用更加灵活，不易出错。

接下来我们继续扩充，介绍一下当关闭页面时需要返回数据，以及上一个页面接收返回数据的具体方法，代码如下。

```
// 关闭页面返回数据
Navigator.pop(context, '返回数据');

// 上一个页面接收返回数据
Navigator.push<String>(context,
        MaterialPageRoute(builder: (BuildContext context) {
      return ButtonSamples(title: '标题', name: '名称');
    })).then((String result) {
      // 处理代码
    });
```

这种路由数据处理情况在很多场景下都会用到。比如，我们注册成功后，返回注册信息给用户中心页面；更改用户头像或信息后返回数据给前一个页面等。大家可以在实际应用中仔细体会它的用法，多实践、多练习。

8.4　生命周期

生命周期就是一个页面对象从创建到销毁的整个状态管理过程。Flutter 中的页面也有生命周期，类似于 Android 中 Activity 的生命周期，不过两者也有很大的不同。在 Flutter 中，生命周期主要体现在 State 的回调函数上。Flutter 生命周期示意图如图 8-1 所示。

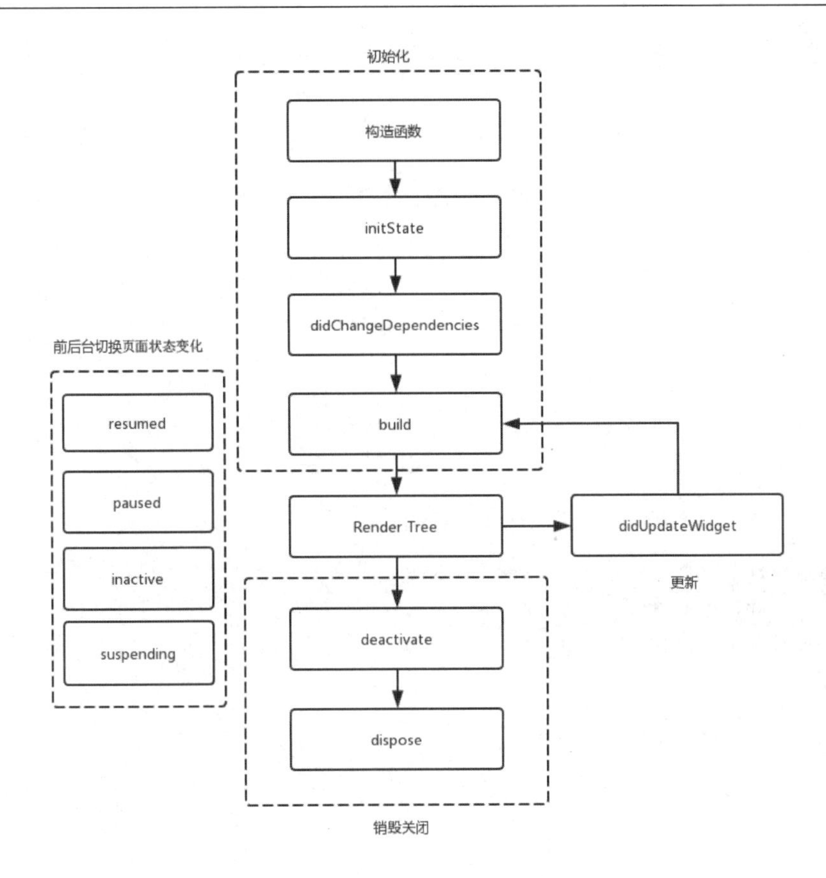

图 8-1　Flutter 生命周期示意图

可以看到，页面在初始化时要按照构造函数 -> initState -> didChangeDependencies -> build 的顺序执行相应操作，然后才会被渲染为一个页面。当页面销毁关闭时，要按照 deactivate -> dispose 顺序执行相应操作。

内部的前后台页面状态变化主要如下。

```
enum AppLifecycleState {
  // 恢复可见
  resumed,
  // 不可见，后台运行，无法处理用户响应
  inactive,
  // 处在并不活动状态，无法处理用户响应
  paused,
  // 应用被立刻暂停
```

```
  suspending,
}
```

当页面更新时，要按照 didUpdateWidget -> build 顺序执行相应操作，并且可能会调用多次。

接下来我们通过代码来看一下 Flutter 页面生命周期中的必要操作。

```
import 'package:flutter/material.dart';

class StateSamples extends StatefulWidget {
  @override
  State<StatefulWidget> createState() {
    return StateSamplesState();
  }
}

class StateSamplesState extends State<StateSamples>
    with WidgetsBindingObserver {
  // 插入渲染树时调用，只调用一次
  @override
  void initState() {
    super.initState();
    WidgetsBinding.instance.addObserver(this);
  }

  // 构建组件时调用
  @override
  Widget build(BuildContext context) {
    return Scaffold(
      appBar: AppBar(
        title: Text('LifeCycleState'),
      ),
      body: Center(
        child: Column(
          children: <Widget>[],
        ),
      ),
    );
  }

  // state 依赖的对象发生变化时调用
```

```
@override
void didChangeDependencies() {
  super.didChangeDependencies();
}

// 组件状态改变时调用，可能会调用多次
@override
void didUpdateWidget(StateSamples oldWidget) {
  super.didUpdateWidget(oldWidget);
}

// 移除渲染树时调用
@override
void deactivate() {
  super.deactivate();
}

// 组件即将销毁时调用
@override
void dispose() {
  super.dispose();
  WidgetsBinding.instance.removeObserver(this);
}

// App 生命周期监听
@override
void didChangeAppLifecycleState(AppLifecycleState state) {
  if (state == AppLifecycleState.resumed) {
    // 恢复可见
  } else if (state == AppLifecycleState.paused) {
    // 处在并不活动状态，无法处理用户响应
  } else if (state == AppLifecycleState.inactive) {
    // 不可见，后台运行，无法处理用户响应
  } else if (state == AppLifecycleState.suspending) {
    // 应用被立刻暂停
  }
  super.didChangeAppLifecycleState(state);
}
```

```
// 其他方法
// 热重载时调用
@override
void reassemble() {
  super.reassemble();
}

// 路由弹出时调用
@override
Future<bool> didPopRoute() {
  return super.didPopRoute();
}

// 有新的路由页面进入时调用
@override
Future<bool> didPushRoute(String route) {
  return super.didPushRoute(route);
}

// 系统窗口配置改变时调用
@override
void didChangeMetrics() {
  super.didChangeMetrics();
}

// 文字缩放时调用
@override
void didChangeTextScaleFactor() {
  super.didChangeTextScaleFactor();
}

// 本地化语言变化时调用
@override
void didChangeLocales(List<Locale> locale) {
  super.didChangeLocales(locale);
}

// 低内存时调用
@override
```

```
  void didHaveMemoryPressure() {
    super.didHaveMemoryPressure();
  }

  // 当前系统辅助功能改变时调用
  @override
  void didChangeAccessibilityFeatures() {
    super.didChangeAccessibilityFeatures();
  }

  // 平台色调主题变化时调用
  @override
  void didChangePlatformBrightness() {
    super.didChangePlatformBrightness();
  }
}
```

通过以上代码，我们可以了解 Flutter 页面从初始化到销毁关闭的整个过程及执行场景。进入一个页面时，会有初始化回调、渲染完毕回调、页面刷新回调、页面销毁回调等一系列回调方法。我们可以利用这些回调方法，配合执行时间点，实现特定需求。

8.5　按键监听

我们知道在 Android 或 iOS 平台上，手机或遥控器上的一些实体按键是可以被监听的，Flutter 上的按键自然也可以被监听。

首先看一下返回键的监听，返回键监听在 Flutter 中比较不一样，需要单独使用一个组件 WillPopScope 来实现。以下示例实现的功能是，在 Flutter 中连续按两次返回键退出应用。

```
class KeyListenerState extends State<KeyListenerSamples> {
  int last = 0;
  int index = 0;

  @override
  void initState() {
    super.initState();
  }

  @override
```

```
  Widget build(BuildContext context) {
    // 要用 WillPopScope 包裹
    return WillPopScope(
      // 编写 onWillPop 逻辑
      onWillPop: _onWillPop,
      child: Scaffold(
          appBar: AppBar(
            title: Text('KeyListener Demo'),
          ),
          body: Center(
            child: Text("按键监听"),
          )),
    );
  }

  // 返回键拦截执行方法
  Future<bool> _onWillPop() {
    int now = DateTime.now().millisecondsSinceEpoch;
    print(now - last);
    if (now - last > 1000) {
      last = now;
      // showToast("再按一次返回键退出");
      return Future.value(false); // 不退出
    } else {
      return Future.value(true); // 退出
    }
  }
}
```

其他按键的监听是使用 RawKeyboardListener 组件来实现的。RawKeyboardListener 继承自 StatefulWidget，属于有状态组件。RawKeyboardListener 的用法如下。

```
class KeyListenerState extends State<KeyListenerSamples> {
  FocusNode focusNode = FocusNode();

  @override
  void initState() {
    super.initState();
    FocusScope.of(context).requestFocus(focusNode);
  }
```

```
@override
Widget build(BuildContext context) {
  return Scaffold(
      appBar: AppBar(
        title: Text('RawKeyboardListener Demo'),
      ),
      // 用 RawKeyboardListener 包裹
      body: RawKeyboardListener(
          // 可以监听的前提是有焦点，我们可以让组件先获取焦点
          focusNode: focusNode,
          onKey: (RawKeyEvent event) {
            // 这里是监听 Android 平台按键，并且是 KeyDown 事件
            if (event is RawKeyDownEvent &&
                event.data is RawKeyEventDataAndroid) {
              RawKeyDownEvent rawKeyDownEvent = event;
              RawKeyEventDataAndroid rawKeyEventDataAndroid =
                  rawKeyDownEvent.data;
              print("keyCode: ${rawKeyEventDataAndroid.keyCode}");
              switch (rawKeyEventDataAndroid.keyCode) {
                // 这里面的 KeyCode 值和 Android 平台的一致
                case 19: // KEY_UP
                  break;
                case 20: // KEY_DOWN
                  break;
                case 21: // KEY_LEFT

                  break;
                case 22: // KEY_RIGHT

                  break;
                case 23: // KEY_CENTER
                  break;
                default:
                  break;
              }
            }
          },
          child: Center(
```

```
              child: Text("按键监听"),
            )),
      );
  }
}
```

当然我们也可以把 RawKeyboardListener 应用在输入框获取和监听上。例如在用户注册时，输入完用户名后按回车键，自动让焦点移动到密码输入框上。希望大家实践以上代码，加深印象，并自己动手尝试编写代码实现其他按键监听功能。

第 9 章

Flutter HTTP 网络请求

HTTP 网络请求是开发语言里比较常用和重要的功能，主要用于实现资源访问、接口数据请求、文件上传、文件下载等操作。Flutter 中 HTTP 网络请求的实现方法主要有三种：通过 io.dart 里的 HttpClient 实现、通过 Dart 原生 HTTP 请求库实现、通过第三方库实现。本章将详细讲解这几种方法的区别和特点，着重介绍前两种方法的使用，并拓展介绍 JSON 编解码、WebSocket 的使用。

9.1 HTTP 网络请求简介

HTTP 网络请求描述了客户端想对指定资源或服务器端执行的操作，主要的请求方式有 GET、POST、HEAD、PUT、DELETE、OPTIONS、TRACE、CONNECT 这 8 种。

1. GET 请求方式

从字面上可以看出，GET 主要用于获取资源。例如通过 URL 从服务器端获取返回的资源时，GET 可以把请求的参数信息拼接在 URL 上，由服务器端进行参数信息解析，然后返回相应的资源给请求者。注意：GET 传递的 URL 数据大小是有限制的，一般限制在 2KB。

2. POST 请求方式

POST 主要用于提交信息、传输信息。POST 请求可以携带很多数据，格式不限，如 JSON、XML、文本等。POST 传递的一些数据和参数不直接拼接在 URL 后面，而放在 HTTP 请求的请求体里，相对 GET 来说比较安全。另外，POST 传递的数据大小是无限制的。

POST 请求包含两部分：请求头（header）和请求体（body）。常见的请求体内容类型有三种：application/x-www-form-urlencoded、application/json、multipart/form-data。

3．HEAD 请求方式

HEAD 主要用于确认 URL 的有效性和资源更新时间，以及查看服务器端状态等。与 GET 不同，HEAD 只需向客户端返回请求头信息，而无须返回请求体内容。

4．PUT 请求方式

PUT 主要用于执行文件传输操作，类似于 FTP 的文件上传一样，请求里包含文件内容，并将文件保存到 URI 指定的服务器端。PUT 和 POST 的主要区别是：对于 PUT 而言，如果前后两个请求相同，后一个请求会把前一个请求覆盖，实现资源修改；而对于 POST，如果前后两个请求相同，后一个请求不会把前一个请求覆盖，实际实现了资源增加。

5．DELETE 请求方式

DELETE 负责告诉服务器端想要删除的资源，执行指定的删除操作。类似于数据库增删改查中的删除操作。

6．OPTIONS 请求方式

OPTIONS 主要用于查询所要请求的 URI 资源服务器端支持的请求方式，即获取指定 URI 支持的客户端提交给服务器端的请求方式。

7．TRACE 请求方式

TRACE 主要用于追踪传输路径，例如我们发起了一个 HTTP 请求，在这个过程中可能会涉及很多路径，TRACE 会告诉服务器端在收到请求后返回一条响应信息，将原始 HTTP 请求信息返回客户端，这样就可以确认在 HTTP 传输过程中请求是否被修改过。

8．CONNECT 请求方式

CONNECT 主要执行连接代理操作。客户端通过 CONNECT 方式与服务器端建立通信隧道，进行 TCP 通信，通过 SSL 和 TLS 安全传输数据。CONNECT 的作用就是告诉服务器端让它代替客户端去请求访问某个资源，然后将数据返回客户端，相当于一个中转媒介。

9.2　实现方式

Flutter 中 HTTP 网络请求的实现方式主要有三种：通过 io.dart 里的 HttpClient 实现、通过 Dart 原生 HTTP 请求库实现、通过第三方库实现。下面我们分别介绍。

9.2.1　通过 io.dart 里的 HttpClient 实现

通过 io.dart 里的 HttpClient 实现的 HTTP 网络请求是非常基础的网络请求，具体实现步骤如下。复杂的网络请求无法通过这种方式实现。

```
import 'dart:convert';
import 'dart:io';

class IOHttpUtils {
  // 创建 HttpClient
  HttpClient _httpClient = HttpClient();

  // 用 async 关键字进行异步请求
  getHttpClient() async {
    _httpClient
        .get('https://abc.com', 8090, '/path1')
        .then((HttpClientRequest request) {
      // 对请求添加 headers 操作，写入请求对象等
      // 调用 close()方法
      return request.close();
    }).then((HttpClientResponse response) {
      // 处理请求响应
      if (response.statusCode == 200) {
        response.transform(utf8.decoder).join().then((String string) {
          print(string);
        });
      } else {
        print("error");
      }
    });
  }

  getUrlHttpClient() async {
```

```
  var url = "https://abc.com:8090/path1";
  _httpClient.getUrl(Uri.parse(url)).then((HttpClientRequest request) {
    // 对请求添加 headers 操作，写入请求对象等
    // 调用 close()方法
    return request.close();
  }).then((HttpClientResponse response) {
    // 处理响应
    if (response.statusCode == 200) {
      response.transform(utf8.decoder).join().then((String string) {
        print(string);
      });
    } else {
      print("error");
    }
  });
}

// POST 请求
postHttpClient() async {
  _httpClient
      .post('https://abc.com', 8090, '/path2')
      .then((HttpClientRequest request) {
    // 这里添加 POST 请求 body 的 ContentType 和内容
    request.headers.contentType = ContentType("application", "json");
    request.write("{\"name\":\"value1\",\"pwd\":\"value2\"}");
    return request.close();
  }).then((HttpClientResponse response) {
    // 处理响应
    if (response.statusCode == 200) {
      response.transform(utf8.decoder).join().then((String string) {
        print(string);
      });
    } else {
      print("error");
    }
  });
}

postUrlHttpClient() async {
```

```
    var url = "https://abc.com:8090/path2";
    _httpClient.postUrl(Uri.parse(url)).then((HttpClientRequest request) {
      // 添加 POST 请求 body 的 ContentType 和内容
      // application 和 x-www-form-urlencoded 数据类型的传输方式
      request.headers.contentType =
          ContentType("application", "x-www-form-urlencoded");
      request.write("name='value1'&pwd='value2'");
      return request.close();
    }).then((HttpClientResponse response) {
      // 处理响应
      if (response.statusCode == 200) {
        response.transform(utf8.decoder).join().then((String string) {
          print(string);
        });
      } else {
        print("error");
      }
    });
  }
}
```

上述代码演示了 HttpClient 的创建过程及基本的 GET 请求方式、POST 请求方式的用法。在组件里请求数据成功后，使用 setState 来更新数据内容和状态即可。总体来看，通过 HttpClient 实现网络请求还是比较简单和方便的。

9.2.2　通过 Dart 原生 HTTP 请求库实现

通过 Dart 原生 HTTP 请求库实现网络请求是比较推荐的，毕竟 Dart 原生的 HTTP 请求库支持的请求方式比较全面，复杂场景也可以实现，如文件上传、文件下载等。

目前 Dart 官方仓库 Dart Pub 里有大量的第三方库和原生库，我们需要在 Dart Pub 里找到原生 HTTP 库并引用。打开 Dart Pub，简单搜索便可以查找到 HTTP 请求库，如图 9-1 所示。

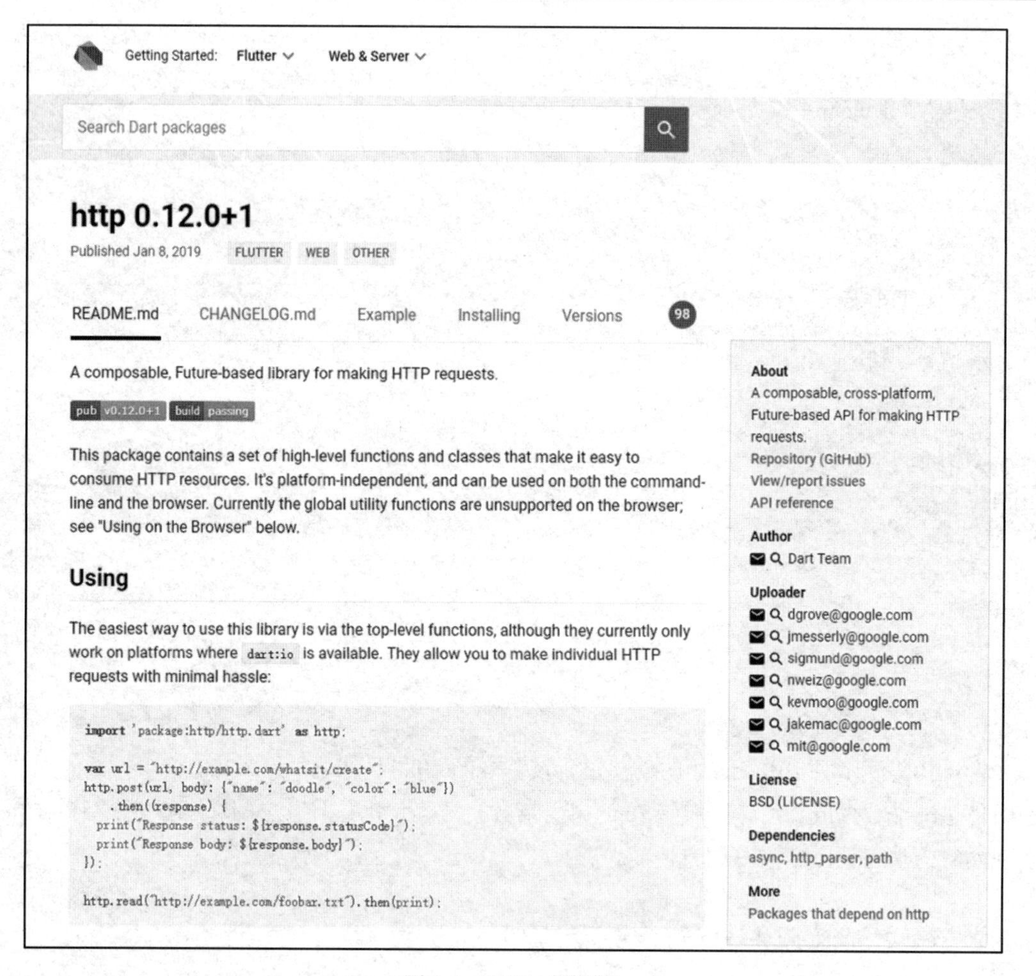

图 9-1　HTTP 请求库

选择【Installing】，查看引用方法，引用 HTTP 请求库，如图 9-2 所示。

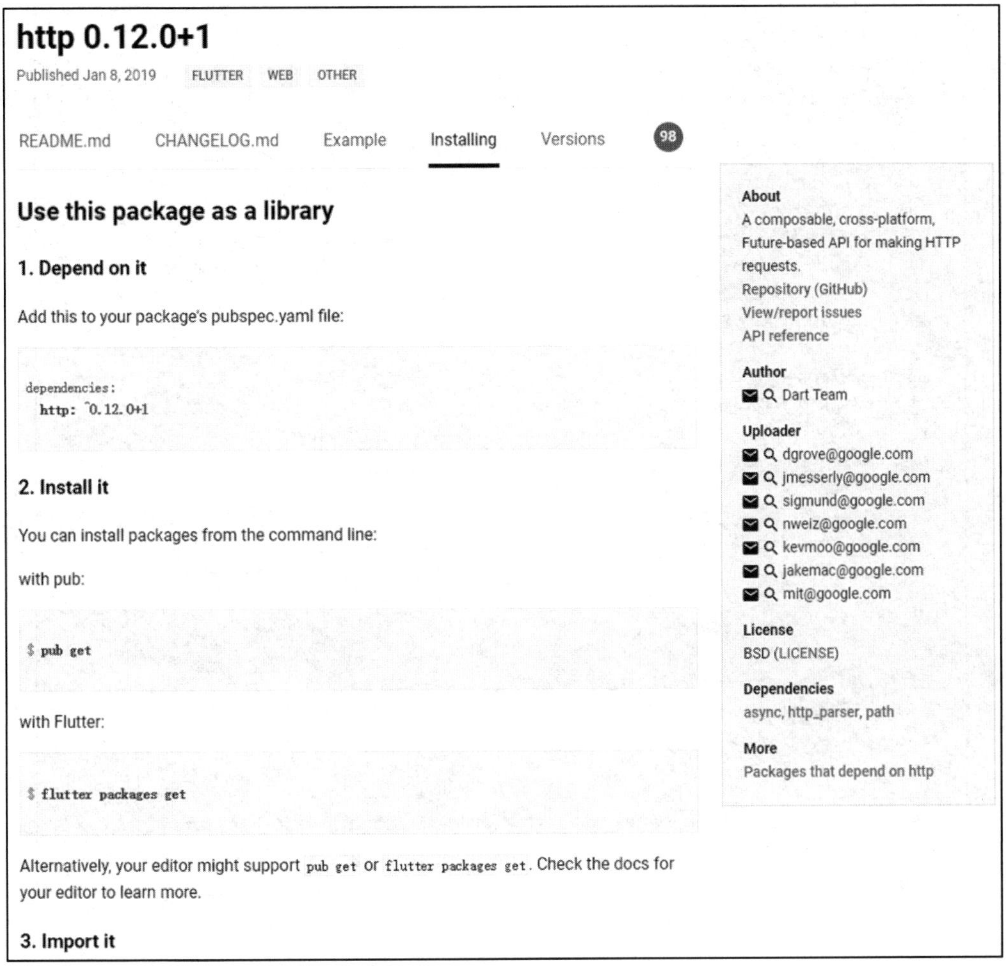

http 0.12.0+1

Published Jan 8, 2019　FLUTTER　WEB　OTHER

README.md　CHANGELOG.md　Example　Installing　Versions　98

Use this package as a library

1. Depend on it

Add this to your package's pubspec.yaml file:

```
dependencies:
  http: ^0.12.0+1
```

2. Install it

You can install packages from the command line:

with pub:

```
$ pub get
```

with Flutter:

```
$ flutter packages get
```

Alternatively, your editor might support `pub get` or `flutter packages get`. Check the docs for your editor to learn more.

3. Import it

About
A composable, cross-platform, Future-based API for making HTTP requests.
Repository (GitHub)
View/report issues
API reference

Author
Dart Team

Uploader
dgrove@google.com
jmesserly@google.com
sigmund@google.com
nweiz@google.com
kevmoo@google.com
jakemac@google.com
mit@google.com

License
BSD (LICENSE)

Dependencies
async, http_parser, path

More
Packages that depend on http

图 9-2　引用 HTTP 请求库

在项目的 pubspec.yaml 配置文件中加入 HTTP 请求库，如图 9-3 所示。

```
! pubspec.yaml ×
1   name: flutter_samples
2   description: A new Flutter project.
3
4   # The following defines the version and build number for your application.
5   # A version number is three numbers separated by dots, like 1.2.43
6   # followed by an optional build number separated by a +.
7   # Both the version and the builder number may be overridden in flutter
8   # build by specifying --build-name and --build-number, respectively.
9   # Read more about versioning at semver.org.
10  version: 1.0.0+1
11
12  environment:
13    sdk: ">=2.0.0-dev.68.0 <3.0.0"
14
15  dependencies:
16    flutter:
17      sdk: flutter
18    http: ^0.12.0+1
19    # The following adds the Cupertino Icons font to your application.
20    # Use with the CupertinoIcons class for iOS style icons.
21    cupertino_icons: ^0.1.2
22
23  dev_dependencies:
24    flutter_test:
25      sdk: flutter
26
27
28  # For information on the generic Dart part of this file, see the
29  # following page: https://www.dartlang.org/tools/pub/pubspec
30
```

图 9-3　在项目的 pubspec.yaml 配置文件中加入 HTTP 请求库

执行完以上操作，就可以在 Dart 文件类里直接引入 HTTP 请求库并使用了。以下是一个完整的通过 Dart 原生 HTTP 请求库实现网络请求的示例。

```
import 'dart:convert';
import 'dart:io';

import 'package:http/http.dart' as http;
import 'package:http_parser/http_parser.dart';

class DartHttpUtils {
  // 创建 client 实例
  var _client = http.Client();
```

```
// 发送 GET 请求
getClient() async {
  var url = "https://abc.com:8090/path1?name=abc&pwd=123";
  _client.get(url).then((http.Response response) {
    // 处理响应信息
    if (response.statusCode == 200) {
      print(response.body);
    } else {
      print('error');
    }
  });
}

// 发送 POST 请求，application 和 x-www-form-urlencoded 数据类型
postUrlencodedClient() async {
  var url = "https://abc.com:8090/path2";
  // 设置 header
  Map<String, String> headersMap = new Map();
  headersMap["content-type"] = "application/x-www-form-urlencoded";
  // 设置 body 参数
  Map<String, String> bodyParams = new Map();
  bodyParams["name"] = "value1";
  bodyParams["pwd"] = "value2";
  _client
      .post(url, headers: headersMap, body: bodyParams, encoding: Utf8Codec())
      .then((http.Response response) {
    if (response.statusCode == 200) {
      print(response.body);
    } else {
      print('error');
    }
  }).catchError((error) {
    print('error');
  });
}

// 发送 POST 请求，application 和 JSON 数据类型
postJsonClient() async {
  var url = "https://abc.com:8090/path3";
```

```
  Map<String, String> headersMap = new Map();
  headersMap["content-type"] = ContentType.json.toString();
  Map<String, String> bodyParams = new Map();
  bodyParams["name"] = "value1";
  bodyParams["pwd"] = "value2";
  _client
      .post(url,
          headers: headersMap,
          body: jsonEncode(bodyParams),
          encoding: Utf8Codec())
      .then((http.Response response) {
    if (response.statusCode == 200) {
      print(response.body);
    } else {
      print('error');
    }
  }).catchError((error) {
    print('error');
  });
}

// 发送 POST 请求，multipart 和 form-data 数据类型
postFormDataClient() async {
  var url = "https://abc.com:8090/path4";
  var client = new http.MultipartRequest("post", Uri.parse(url));
  client.fields["name"] = "value1";
  client.fields["pwd"] = "value2";
  client.send().then((http.StreamedResponse response) {
    if (response.statusCode == 200) {
      response.stream.transform(utf8.decoder).join().then((String string) {
        print(string);
      });
    } else {
      print('error');
    }
  }).catchError((error) {
    print('error');
  });
}
```

```
// 发送 POST 请求，上传文件
  postFileClient() async {
    var url = "https://abc.com:8090/path5";
    var client = new http.MultipartRequest("post", Uri.parse(url));
    http.MultipartFile.fromPath('file', 'sdcard/img.png',
            filename: 'img.png', contentType: MediaType('image', 'png'))
        .then((http.MultipartFile file) {
      client.files.add(file);
      client.fields["description"] = "descriptiondescription";
      client.send().then((http.StreamedResponse response) {
        if (response.statusCode == 200) {
          response.stream.transform(utf8.decoder).join().then((String string) {
            print(string);
          });
        } else {
          response.stream.transform(utf8.decoder).join().then((String string) {
            print(string);
          });
        }
      }).catchError((error) {
        print(error);
      });
    });
  }
}
```

通过以上代码可以看出，原生 HTTP 请求库可以实现所有通用和复杂场景下的 GET 请求和 POST 请求，并且使用方便、配置简单。其余的 HEAD、PUT、DELETE 等请求的实现方法类似，大家可以自行尝试。在组件里请求数据成功后，使用 setState 可以更新数据内容和状态。

9.2.3　通过第三方库实现

有很多 Flutter 第三方库可用于实现 HTTP 网络请求，其中比较典型的有国内开发者开发的 dio 库，dio 库支持多文件上传、下载、并发请求等复杂场景。在 Dart Pub 中可以搜索 dio 库，如图 9-4 所示。

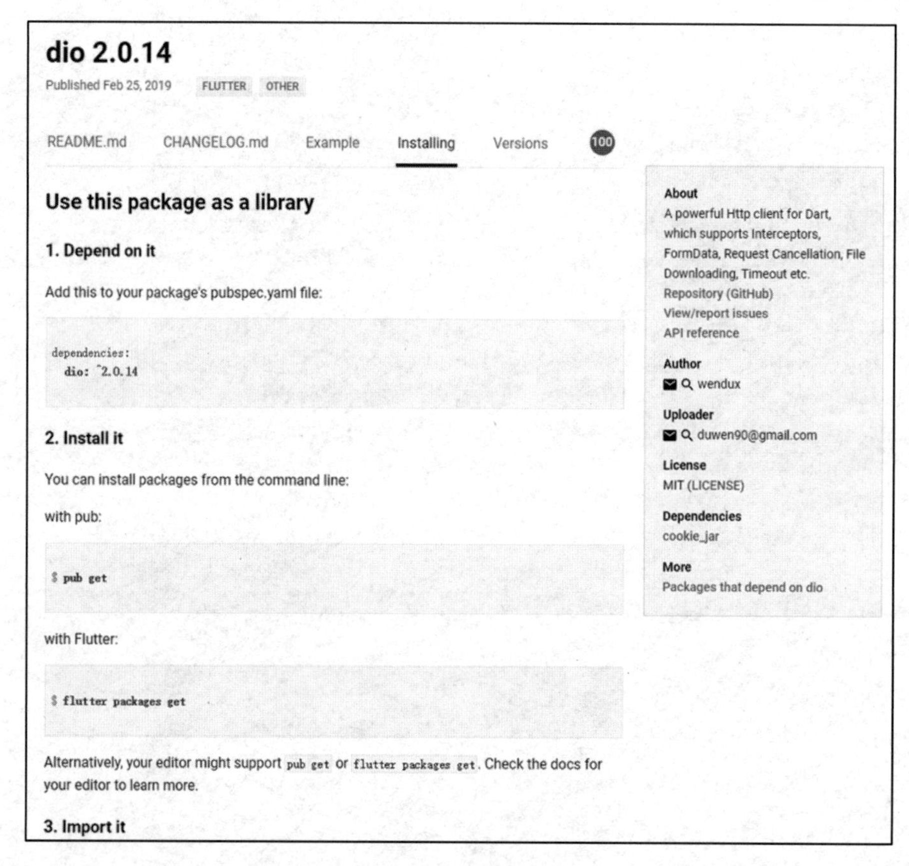

图 9-4　dio 库

在实际应用中，要想引用 dio 库的 API 来实现 HTTP 网络请求，需要在项目的 pubspec.yaml 配置文件里加入如下内容。

```
dependencies:
  dio: ^2.0.14
```

完整的 dio 库的用法示例如下。

```
import 'dart:io';

import 'package:dio/dio.dart';

class DartHttpUtils {
  // 通过 BaseOptions 配置 dio 库
  Dio _dio = Dio(BaseOptions(
      baseUrl: "https://abc.com:8090/",
```

```
    connectTimeout: 5000,
    receiveTimeout: 5000));

// dio 库的 GET 请求
getDio() async {
  var url = "/path1?name=abc&pwd=123";
  _dio.get(url).then((Response response) {
    if (response.statusCode == 200) {
      print(response.data.toString());
    }
  });
}

getUriDio() async {
  var url = "/path1?name=abc&pwd=123";
  _dio.getUri(Uri.parse(url)).then((Response response) {
    if (response.statusCode == 200) {
      print(response.data.toString());
    }
  }).catchError((error) {
    print(error.toString());
  });
}

// dio 库的 GET 请求，通过 queryParameters 配置传递参数
getParametersDio() async {
  var url = "/path1";
  _dio.get(url, queryParameters: {"name": 'abc', "pwd": 123}).then(
      (Response response) {
    if (response.statusCode == 200) {
      print(response.data.toString());
    }
  }).catchError((error) {
    print(error.toString());
  });
}

// 发送 POST 请求，application 和 x-www-form-urlencoded 数据类型
postUrlencodedDio() async {
```

```
  var url = "/path2";
  _dio
      .post(url,
          data: {"name": 'value1', "pwd": 123},
          options: Options(
              contentType:
                  ContentType.parse("application/x-www-form-urlencoded")))
      .then((Response response) {
    if (response.statusCode == 200) {
      print(response.data.toString());
    }
  }).catchError((error) {
    print(error.toString());
  });
}

// 发送 POST 请求，application 和 JSON 数据类型
postJsonDio() async {
  var url = "/path3";
  _dio
      .post(url,
          data: {"name": 'value1', "pwd": 123},
          options: Options(contentType: ContentType.json))
      .then((Response response) {
    if (response.statusCode == 200) {
      print(response.data.toString());
    }
  }).catchError((error) {
    print(error.toString());
  });
}

// 发送 POST 请求，multipart 和 form-data 数据类型
postFormDataDio() async {
  var url = "/path4";
  FormData _formData = FormData.from({
    "name": "value1",
    "pwd": 123,
  });
```

```
    _dio.post(url, data: _formData).then((Response response) {
      if (response.statusCode == 200) {
        print(response.data.toString());
      }
    }).catchError((error) {
      print(error.toString());
    });
  }

  // 发送 POST 请求，上传文件
  postFileDio() async {
    var url = "/path5";
    FormData _formData = FormData.from({
      "description": "descriptiondescription",
      "file": UploadFileInfo(File("./example/upload.txt"), "upload.txt")
    });
    _dio.post(url, data: _formData).then((Response response) {
      if (response.statusCode == 200) {
        print(response.data.toString());
      }
    }).catchError((error) {
      print(error.toString());
    });
  }

  // 下载文件
  downloadFileDio() {
    var urlPath = "https://abc.com:8090/";
    var savePath = "./abc.html";
    _dio.download(urlPath, savePath).then((Response response) {
      if (response.statusCode == 200) {
        print(response.data.toString());
      }
    }).catchError((error) {
      print(error.toString());
    });
  }
}
```

通过以上代码可以看出，dio 库的风格和逻辑非常清晰，可以实现复杂的 HTTP 网络请求。

在上面介绍的三种实现 HTTP 请求的方法中，推荐使用第二种方法和第三种方法。因为这两种方法功能比较全面，能实现更复杂的网络请求场景，而且代码逻辑清晰，不容易出错。大家可以根据自己的习惯和实际需求来合理选择。

9.3 Flutter JSON 编解码

在实际开发中，经常会用到数据交换格式，如 JSON、XML。Flutter 同样可以处理 JSON 格式编解码操作，我们可以在 Flutter 中实现将一个 JSON 字符串转为实体对象或将一个实体对象转为 JSON 字符串的操作。本节主要讲解 Flutter 里的 JSON 编解码用法、JSON 编解码优化、JSON 自动序列化解码。

9.3.1 JSON 编解码用法详解

当请求网络数据接口或缓存某些结构数据时，一般都会用到 JSON 格式。JSON 在移动端、后端、前端开发中应用都非常广泛。在 Flutter 中，解析 JSON 格式要使用 dart:convert 里的函数类。

我们先来看一个最简单的 JSON 解码和编码的例子，如下。

```
// JSON 解码
// 定义一个 JSON 格式字符串
String _jsonString = '{"name": "Flutter Book","author": "Google"}';
// 使用 json.decode 进行解码
Map<String, dynamic> book = json.decode(_jsonString);
// 解码后调用获取值
Column(
    children: <Widget>[
        Text('Book Name: ${book['name']}'),
        Text('Book Author: ${book['author']}'),
    ],
));

// JSON 编码
// 使用 json.encode 将实体对象编码为 JSON 字符串
String _bookJson = json.encode(book);
```

以上是一个最简单的例子，实际开发中可能会遇到更加复杂的嵌套 JSON 结构。以下是一个复杂一些的 JSON 解析的例子。

```
String _jsonListString =
    '[{"name": "Flutter Book","author": "Google"},{"name": "Dart Book","author": "G
oogle"}]';
// 解析成 List
List books = json.decode(_jsonListString);
// 调用取值
print(books[0]["name"]);
```

通过以上代码可以看出，Flutter 的 JSON 编解码方式和其他平台大同小异，只要能够区分出 JSON 字符串的数据结构即可按照实际需求解析成对象或对象数组集合。大家可以找一些常用的 JSON 接口进行实践，加深理解。

9.3.2　JSON 编解码优化

接下来我们优化一下 JSON 编解码过程，以便使用时更加高效、便捷。首先定义一个实体对象 Book，代码如下。

```
class Book {
  final String name;
  final String author;

  Book(this.name, this.author);

  Book.fromJson(Map<String, dynamic> json)
      : name = json['name'],
        author = json['author'];

  Map<String, dynamic> toJson() => {
        'name': name,
        'author': author,
      };
}
```

可以看到，实体对象 Book 里新增了两个方法：fromJson 和 toJson。这两个方法分别用于将 JSON 字符串解码为实体对象，以及将实体对象转为 JSON 字符串。通过调用 fromJson 方法和 toJson 方法实现 JSON 的编解码十分清晰、高效。这种方法扩展性强、安全，也能体现面向对象编程的封装性，降低代码耦合度，具体调用方法如下。

```
// 解码 JSON 字符串
Map<String, dynamic> bookMap = json.decode(_jsonString);
// 调用 Book 里的 fromJson 方法，转换创建实体对象 Book
var bookBean = Book.fromJson(bookMap);
// 具体使用方法
print(bookBean.name);
print(bookBean.author);

// 编码 JSON 字符串
// 直接调用，推荐
String beanString = json.encode(bookBean);
// 调用实体对象的 toJson 方法，不推荐
String beanString = json.encode(bookBean.toJson());
```

9.3.3　JSON 自动序列化解码

在实际开发中，实体对象的属性可能有很多，并且内部嵌套对象或 List 集合。对于这种更加复杂的对象，如果我们依然一个一个编写方法，进行属性赋值、JSON 编解码，那将非常耗费时间。为了应对这种情况，Flutter 官方提供了一个插件库 json_serializable。json_serializable 可以帮助我们实现 JSON 自动序列化解码，免去很多重复的工作，避免出错。

要想使用 json_serializable 插件库，首先需要在 pubspec.yaml 里添加依赖，如下。

```
dependencies:
  json_serializable: ^2.3.0
```

运行 flutter packages get 命令，或者在 Visual Studio 里更改配置文件，保存后会自动同步相关资源。这里建议使用 Flutter SDK 最新版本。然后，在需要使用的地方导入 JSON 自动序列化类，如下。

```
import 'package:json_serializable/json_serializable.dart';
```

完成以上两步，我们就可以使用 json_serializable 插件库了。json_serializable 的简单用法如下。

```
// 新建一个实体对象
import 'package:json_annotation/json_annotation.dart';
// book.g.dart 是自动生成的名字
part 'book.g.dart';
// JSON 序列化注解
@JsonSerializable(nullable: false)
```

```
class Book{
  Book(this.name, this.author);

  String name;
  String author;

  factory Book.fromJson(Map<String, dynamic> json) => _$BookFromJson(json);
  Map<String, dynamic> toJson() => _$BookToJson(this);
}

// 配置自动生成命令
flutter packages pub run build_runner build
// 或使用以下命令，每次新增、修改实体对象后都会自动监听
flutter packages pub run build_runner watch
// 执行完命令后，会在同目录文件夹生成对应实体对象

// book.g.dart
// 自动生成的代码
// 自动完成属性映射工作
// GENERATED CODE - DO NOT MODIFY BY HAND

part of 'book.dart';

// **************************************************************************
// JsonSerializableGenerator
// **************************************************************************

Book _$BookFromJson(Map<String, dynamic> json) {
  return Book(json['name'] as String, json['author'] as String);
}

Map<String, dynamic> _$BookToJson(Book instance) =>
<String, dynamic>{'name': instance.name, 'author': instance.author};
```

通过以上代码，Book 就变成了由 book.dart 和 book.g.dart 两部分组成的对象。如果将这两部分完整放在一起，代码如下。

```
class Book{
  Book(this.name, this.author);
```

```
  String name;
  String author;

  Book _$BookFromJson(Map<String, dynamic> json) {
    return Book(json['name'] as String, json['author'] as String);
  }

  Map<String, dynamic> _$BookToJson(Book instance) =>
      <String, dynamic>{'name': instance.name, 'author': instance.author};
  }
```

自动生成映射方法后，我们就可以调用 BookFromJson 和 BookToJson 来实现 JSON 编解码操作了。具体使用方法和前面介绍的基本没有区别，代码如下。

```
// JSON 字符串
String _jsonString = '{"name": "Flutter Book","author": "Google"}';
// 解码成实体对象
Map<String, dynamic> bookMap = json.decode(_jsonString);
var bookBean = Book.fromJson(bookMap);
// 调用
print(bookBean.name);
print(bookBean.author);
// 将实体对象编码为 JSON 字符串
String beanString = json.encode(bookBean);
```

以上几种方式是Flutter对JSON字符串执行的编解码操作，借助这些操作，我们可以在JSON字符串和实体对象之间来回转换，实现实际项目需求。

9.4 Flutter WebSocket 的使用

WebSocket 在推送、聊天、数据传输场景中经常使用，Flutter SDK 自带 WebSocket 功能，通过第三方插件库也可以实现WebSocket功能。本节我们将介绍WebSocket基本知识、WebSocket简单用法，以及通过第三方插件库进行 WebSocket 通信的方法。

9.4.1 WebSocket 简介

提起 Socket、TCP、UDP，相信大家并不陌生，它们都可以用来进行数据通信。WebSocket 也是基于 TCP 实现的全双工通信协议，它可以实现客户端和服务器端数据双向传输。只需连接一次，便可以持久化通信。

我们知道 HTTP 是单向请求/响应式协议，是无连接的协议。当客户端发起请求时，服务器端将对请求做出相应处理，如果用 HTTP 来实现推送，当服务器端有数据或状态变化时，要不停地发送 HTTP 请求给服务器端进行轮询，效率会比较低，容易出错，而且浪费资源。

面对以上问题，我们可以用 WebSocket 来解决。使用 WebSocket 很简单，一般包括连接 WebSocket 服务器、发送消息、接收消息、关闭 WebSocket 连接这几步。

9.4.2 WebSocket 基本用法

目前 Flutter SDK 中已经自带 WebSocket API。按照连接 WebSocket 服务器、发送消息、接收消息、关闭 WebSocket 连接几个步骤，使用 WebSocket 的具体代码如下。

```
// 导入 WebSocket 包
import 'dart:io';
...
// 连接 WebSocket 服务器
Future<WebSocket> webSocketFuture =
        WebSocket.connect('ws://192.168.1.8:8080');

// WebSocket.connect 返回的是 Future<WebSocket>对象
static WebSocket _webSocket;

webSocketFuture.then((WebSocket ws) {
      _webSocket = ws;

      void onData(dynamic content) {
        print('收到');
      }
      // 调用 add 方法发送消息
      _webSocket.add('message');
      // 调用 listen 方法监听接收消息
     _webSocket.listen(onData, onDone: () {
        print('onDone');
      }, onError: () {
        print('onError');
      }, cancelOnError: true);
```

```
    });
...
// 发送消息
_webSocket.add('发送消息内容');
...
// 调用 listen 方法监听接收消息
void onData(dynamic content) {
        print('收到消息:'+content);
    }
_webSocket.listen(onData, onDone: () {
        print('onDone');
      }, onError: () {
        print('onError');
      }, cancelOnError: true);
...
// 关闭 WebSocket 连接
_webSocket.close();
```

怎么样，是不是很简单。接下来我们介绍通过第三方插件库进行 WebSocket 通信的方法。

9.4.3　通过第三方插件库进行 WebSocket 通信

本节我们选择的第三方插件是 web_socket_channel。基本实现步骤也是连接 WebSocket 服务器、发送消息、接收消息、关闭 WebSocket 连接。首先要在项目的 pubspec.yaml 里加入引用，代码如下。

```
dependencies:
  web_socket_channel: ^1.0.13
```

加入引用后，按照上述顺序实现 WebSocket 通信，代码如下。

```
// 导入相关包
import 'package:web_socket_channel/io.dart';
import 'package:web_socket_channel/status.dart' as status;
...
// 连接 WebSocket 服务器
var channel = IOWebSocketChannel.connect("ws://192.168.1.8:8080");
...
// 通过 IOWebSocketChannel 进行各种操作
// 发送消息
channel.sink.add("connected!");
```

```
...
// 监听、接收消息
channel.stream.listen((message) {
      print('收到消息:' + message);
});
...
// 关闭 WebSocket 连接
channel.sink.close();
```

以上是 通过第三方插件库实现 WebSocket 通信功能的代码。当然 Flutter 也支持 Socket 相关 API 操作，大家可以自行学习。

第 10 章

Flutter 文件操作与数据库操作

实际开发中离不开文件操作，例如缓存数据、创建和删除文件、读取数据、加载图片等。Flutter 也提供了相关的文件操作 API，同时也支持数据库操作。本章将结合具体示例介绍 Flutter 的文件操作、数据库操作，同时介绍 Flutter 手势的用法。

10.1 文件操作

对于 Flutter 的文件操作，我们可以使用原生 API 来实现，主要涉及 File 和 Directory 这两个组件，也可以使用第三方库来实现，同时还可以自己编写插件库来实现。本节主要介绍使用 Flutter 原生 API 和 Flutter 第三方插件库来实现文件操作的方法。

以下是使用 Flutter 原生 API 实现文件操作的方法，具体如下。

```
// File 和 Directory 操作都在 dart:io 包里
import 'dart:io';

/// 创建文件夹
Future directory1() async {
  var directory = Directory("temp1");
  directory.create();
}

/// 递归创建文件夹
void directory2() {
  Directory('dir/subdir').create(recursive: true).then((Directory directory) {
    print(directory.path);
  });
```

```
}

/// 列出目录文件
void directory3() {
  var systemTempDir = Directory('sdcard');
  systemTempDir
      .list(recursive: true, followLinks: false)
      .listen((FileSystemEntity entity) {
    print(entity.path);
  });
}

/// 判断文件夹是否存在
void directory4() {
  var directory = Directory('dir/subdir');
  directory.exists().then((isThere) {
    print(isThere);
  });
}

/// 文件夹重命名
void directory5() {
  var directory = Directory('dir/subdir');
  directory.rename('dir/subRightDir').then((Directory directory) {});
}

/// 判断文件是否存在
void file1() {
  var file = File('E:/file.txt');
  file.exists().then((isThere) {
    print(isThere);
  });
}

/// 创建文件
void file2() {
  var file = File('E:/file2.txt');
  file.create().then((file) {
    print(file);
  });
}
```

```
/// 最强大的读取文件方式：Stream 方式
void file3() {
  var file = File('E:/file.txt');
  Stream<List<int>> inputStream = file.openRead();
  inputStream
      .transform(utf8.decoder)
      .transform(LineSplitter())
      .listen((string) {
    print(string);
  }).onDone(() {
    print('File is now closed.');
  });
}

/// 向文件写入内容，方式 1
void file4() {
  var file = File('E:/file2.txt');
  file.writeAsString('some content').then((file) {
    print(file);
  });
}

/// 向文件写入内容，方式 2
void file5() {
  var file = File('E:/file2.txt');
  var sink = file.openWrite();
  sink.write('FILE ACCESSED ${new DateTime.now()}\n');
  sink.close();
}

/// 获取文件长度等信息
void file6() {
  var file = File('E:/file.txt');
  file.length().then((len) {
    print(len);
  });
}
```

通过上面的代码可以看到，使用 Flutter 原生 API 进行文件操作的核心就是通过实例化 File

和 Directory 这两个组件来调用各种操作方法。原理非常简单，容易理解，这里不做过多解释，大家可以自己动手实践。

接下来我们再看一下通过第三方插件库来实现文件操作的方法。Flutter 官方提供了一个文件操作插件库 path_provider。使用时，需要先在 pubspec.yaml 文件里添加引用。

```
dependencies:
  path_provider: ^1.1.0
```

接下来在使用的地方导入对应的文件操作类 path_provider.dart。

```
import 'package:path_provider/path_provider.dart';
```

以获取缓存目录、获取应用数据目录、获取设备存储根目录为例，实现文件操作的具体代码如下。

```
// 获取缓存目录，相当于 Android 的 getCacheDir 和 iOS 的 NSCachesDirectory
void getCacheDir() {
  Future<Directory> tempDir = getTemporaryDirectory();
  tempDir.then((Directory directory) {
    print(directory.path);
  });
}

// 获取应用数据目录，相当于 Android 的 getDataDirectory 和 iOS 的 NSDocumentDirectory
void getAppDocDir() {
  Future<Directory> appDocDir = getApplicationDocumentsDirectory();
  appDocDir.then((Directory directory) {
    print(directory.path);
  });
}

// 获取设备存储根目录，相当于 Android 的 getExternalStorageDirectory
void getAppExternalDir() {
  Future<Directory> appExternalDir = getExternalStorageDirectory();
  appExternalDir.then((Directory directory) {
    print(directory.path);
  });
}}
```

通过以上代码可以看到，path_provider 插件库的主要功能是将手机内部的路径进行转化，转化为 File 和 Directory 组件可以直接使用的路径。这样一来，我们就可以非常方便地进行

Android 和 iOS 等平台的文件操作。

接下来我们来看一下 Flutter 中的图片操作。在前面的章节中，我们介绍过 Image 组件的基本用法，涉及图片加载和显示操作。这里我们将结合 File 组件介绍拓展操作。

Flutter 中图片的加载源一般有网络图片加载、本地图片加载、资源文件里的图片加载这几种。网络加载图片的代码如下。

```
// 直接调用 network 方法加载
Image.network(imageUrl)

// 通过 NetworkImage 方法加载，需要 Image 组件包裹
Image(
  image: NetworkImage(imageUrl),
)

// 在加载网络图片时放置一个加载成功前的占位图
FadeInImage.assetNetwork(
  placeholder: "assets/flutter-mark-square-64.png",
  image: imageUrl,
)
```

我们还可以在加载网络图片时进行一些个性化配置，在以上代码组件中配置相应的属性即可实现。接下来我们来看一下使用 Flutter 加载本地图片的方法，代码如下。

```
// 直接调用 file 方法从硬盘加载读取的图片
Image.file(
  File('/sdcard/img.png'),
  width: 200,
  height: 80,
)

// 调用 FileImage 组件加载本地图片
Image(
  image: FileImage(File('/sdcard/img.png')),
)

// 从内存、缓存中读取 byte 数组图片
Image.memory(bytes)

Image(
```

```
  image: MemoryImage(bytes),
)
```

同样，我们在加载本地图片时也可以进行个性化配置，如图片显示方式、尺寸，如果遇到不同平台路径适配问题，可以利用之前讲过的 path_provider 插件库来解决。最后我们来看一下从资源文件里加载图片的方法，代码如下。

```
// 从项目目录里读取图片，需要在 pubspec.yaml 中注册路径
assets:
    - assets/flutter-mark-square-64.png
...
// assets 目录是项目的根目录
// 直接调用 Image 组件的 asset 方法加载
Image.asset("assets/flutter-mark-square-64.png"),

// 使用 AssetImage 方法加载图片
Image(
  image: AssetImage("assets/flutter-mark-square-64.png"),
)
```

总体来说，Flutter 中的图片加载操作很简单。大家可以结合 Image 组件进行实践，并对加载的图片做一些个性化配置。

10.2　手势操作

在移动应用开发中，经常会用到手势操作，这也是移动平台应用独有的操作特性。我们可以使用手势进行密码设置与解锁，定义一系列重要操作等。Flutter 同样支持手势操作，本节将扩展讲解。

Flutter 的手势包括单击、双击、长按、拖曳、移动、缩放等，这些手势操作都是通过 GestureDetector 组件来实现的。GestureDetector 继承自 StatelessWidget，属于无状态组件。它的构造方法和属性参数含义如下。

```
GestureDetector({
    Key key,
    // 子组件
    this.child,
    // 单击
    this.onTap,
    // 双击
```

```
this.onDoubleTap,
// 长按
this.onLongPress,
// 垂直拖曳按下
this.onVerticalDragDown,
// 垂直拖曳开始
this.onVerticalDragStart,
this.onVerticalDragUpdate,
// 垂直拖曳结束
this.onVerticalDragEnd,
// 垂直拖曳取消
this.onVerticalDragCancel,
// 在水平方向执行上述操作
this.onHorizontalDragDown,
this.onHorizontalDragStart,
this.onHorizontalDragUpdate,
this.onHorizontalDragEnd,
this.onHorizontalDragCancel,
this.onForcePressStart,
this.onForcePressPeak,
this.onForcePressUpdate,
this.onForcePressEnd,
// 手指按下
this.onPanDown,
// 手指按下开始滑动
this.onPanStart,
// 手指按下滑动过程中
this.onPanUpdate,
// 手指抬起
this.onPanEnd,
// 滑动取消
this.onPanCancel,
// 缩放开始
this.onScaleStart,
// 缩放中
this.onScaleUpdate,
// 缩放结束
this.onScaleEnd,
this.behavior,
```

```
    this.excludeFromSemantics = false,
    this.dragStartBehavior = DragStartBehavior.start,
  })
```

通过以上构造方法和属性参数可以看到，GestureDetector 组件基本上可以实现大部分手势的监听和处理。接下来我们来看一下 GestureDetector 中几个常用、重要方法（单击、双击、长按事件）的具体用法。代码如下。

```
var gestureStatus = 'Gesture';
...
Center(
      child: GestureDetector(
        child: RaisedButton(
          child: Text(gestureStatus),
        ),
        // 单击
        onTap: () {
          setState(() {
            gestureStatus = 'onTap';
          });
        },
        // 双击
        onDoubleTap: () {
          setState(() {
            gestureStatus = 'onDoubleTap';
          });
        },
        // 长按事件
        onLongPress: () {
          setState(() {
            gestureStatus = 'onLongPress';
          });
        },
      ),
    )
```

我们只要在需要添加手势的组件外面包裹一个 GestureDetector 组件，然后在里面实现对应的手势操作方法即可。接下来我们再看一下使用 GestureDetector 组件来对垂直方向拖曳进行监听的方法，代码如下。

```
// Flutter 支持沿某个方向进行监听
```

```
// 垂直方向
class GestureSamplesState extends State<GestureSamples> {
  var gestureStatus = 'Gesture';
  var _top = 100.0;

  @override
  void initState() {
    super.initState();
  }

  @override
  Widget build(BuildContext context) {
    return getstureVertical(context);
  }

  Widget getstureVertical(BuildContext context) {
    return Scaffold(
      appBar: AppBar(
        title: Text('Gesture'),
      ),
      body: Container(
          child: Stack(
        children: <Widget>[
          Positioned(
            top: _top,
            left: 100.0,
            child: GestureDetector(
                child: Icon(Icons.history),
                // 垂直方向拖曳监听
                onVerticalDragUpdate: (DragUpdateDetails details) {
                  setState(() {
                    // 动态更新垂直坐标
                    _top += details.delta.dy;
                  });
                }),
          )
        ],
      )),
    );
```

```
  }

}
```

对水平方向拖曳进行监听的方法同理可得。如果一个组件既有水平方向拖曳监听又有垂直方向拖曳监听，执行哪个方向的操作主要取决于手指在哪个方向上的滑动位移大。除了以上几个操作， GestureDetector 组件还可以实现缩放监听操作，代码如下。

```
class GestureSamplesState extends State<GestureSamples> {
  var gestureStatus = 'Gesture';

  var _width = 180.0;
  var _scale = 1.0;

  @override
  void initState() {
    super.initState();
  }

  @override
  Widget build(BuildContext context) {
    return getstureScale(context);
  }

  Widget getstureScale(BuildContext context) {
    return Scaffold(
      appBar: AppBar(
        title: Text('Gesture'),
      ),
      body: Container(
          child: Stack(
        children: <Widget>[
          Center(
            child: GestureDetector(
              // 动态控制宽度，高度自适应
              child: Image.asset(
                "assets/image_appbar.jpg",
                width: _width,
              ),
              onScaleStart: (ScaleStartDetails details) {
```

```
              print('onScaleStart:$details.focalPoint');
            },
            onScaleEnd: (ScaleEndDetails details) {
              print('onScaleEnd:$details.focalPoint');
            },
            onScaleUpdate: (ScaleUpdateDetails details) {
              setState(() {
                // 除了缩放也可以进行旋转
                // details.rotation.clamp(0, 360);
                // 缩放比例是 50%~500%
                _scale = details.scale.clamp(0.5, 5);
                _width = 180 * _scale;
              });
            },
          ),
        )
      ],
    )),
  );
 }
}
```

通过以上的介绍，大家应该对 Flutter 的手势操作有了大致的了解。我们可以通过 Flutter 的手势操作实现很多功能，如图片缩放、组件拖曳、控制监听等。针对手势监听功能，大家可以关注 Listener、GestureRecognizer、RawGestureDetector 的用法，对比着进行扩展学习。

10.3　数据库操作

Flutter 原生 API 是不支持数据库操作的，因此要以插件库方式进行数据库操作。Flutter 官方提供了数据库插件库 sqflite，通过 sqflite 可实现数据库的增删改查操作。使用 sqflite 的时候，需要先在 pubspec.yaml 里添加依赖。

```
dependencies:
  sqflite: ^1.1.5
```

然后在使用的地方导入 sqflite.dart 类。

```
import 'package:sqflite/sqflite.dart';
```

完成以上操作就可以实际应用 sqflite 了。首先我们要建立一个数据库表对应的实体对象，这里命名为Note，该对象的属性和相关方法如下。

```
final String tableName = 'notes';// 数据库表名
final String columnId = '_id';// 属性名，记录数据
final String columnTitle = 'title';// 属性名，记录标题
final String columnContent = 'content';// 属性名，记录内容

// 实体对象
class Note {
  int id;
  String title;
  String content;

  // 将实体对象转为数据集合
  Map<String, dynamic> toMap() {
    var map = <String, dynamic>{
      columnTitle: title,
      columnContent: content,
    };
    if (id != null) {
      map[columnId] = id;
    }
    return map;
  }

  // 构造方法/实例化方法
  Note();

  // 通过数据集合返回一个实体对象
  Note.fromMap(Map<String, dynamic> map) {
    id = map[columnId];
    title = map[columnTitle];
    content = map[columnContent];
  }
}
```

通过以上代码可以看到，Note实体对象里不仅有实体的属性，还包含了一些将数据和对象相互转换的方法。接下来需要创建数据库表的操作工具类NoteDbHelper，代码如下。

```
import 'package:sqflite/sqflite.dart';
```

```
import 'Note.dart';

// 数据库表操作工具类
class NoteDbHelper {
  Database db;

  Future open(String path) async {
    // 打开/创建数据库
    db = await openDatabase(path, version: 1,
        onCreate: (Database db, int version) async {
      await db.execute('''
create table Notes (
  _id integer primary key autoincrement,
  title text not null,
  content text not null)
''');
    });
  }

  Database getDatabase(){
    return db;
  }

  // 增加一条数据
  Future<Note> insert(Note note) async {
    note.id = await db.insert("notes", note.toMap());
    return note;
  }

  // 通过 ID 查询一条数据
  Future<Note> getNoteById(int id) async {
    List<Map> maps = await db.query('notes',
        columns: [columnId, columnTitle, columnContent],
        where: '_id = ?',
        whereArgs: [id]);
    if (maps.length > 0) {
      return Note.fromMap(maps.first);
    }
    return null;
```

```
  }

  // 通过 ID 删除一条数据
  Future<int> deleteById(int id) async {
    return await db.delete('notes', where: '_id = ?', whereArgs: [id]);
  }

  // 更新数据
  Future<int> update(Note note) async {
    return await db.update('notes', note.toMap(),
        where: '_id = ?', whereArgs: [note.id]);
  }

  // 关闭数据库
  Future close() async => db.close();
}
```

从上面的代码可以看出，NoteDbHelper 工具类的结构和 Android 中的数据库操作工具类很像，都是由打开数据库、数据库增删改查、关闭数据库几个核心部分组成的。最后我们来看一下如何调用写好的工具类，代码如下。

```
import 'package:flutter/material.dart';
import 'package:flutter/widgets.dart';
import 'package:flutter_samples/samples/note_db_helper.dart';
import 'package:path/path.dart';
import 'package:sqflite/sqflite.dart';

import 'Note.dart';
...
class SQLiteSamplesState extends State<SQLiteSamples> {
  NoteDbHelper noteDbHelpter;

  @override
  void initState() {
super.initState();
    // 实例化工具类
    noteDbHelpter = NoteDbHelper();
    String databasesPath = getDatabasesPath().toString();
    String path = join(databasesPath, 'notesDb.db');
    // 打开数据库
```

```
    noteDbHelpter.open(path);
    // 调用查询方法
    noteDbHelpter.getNoteById(1).then((Note note) {
      print(note.title);
    });

    // 也可以这样使用
    noteDbHelpter
        .getDatabase()
        .query('notes')
        .then((List<Map<String, dynamic>> records) {
      Map<String, dynamic> mapRead = records.first;
      // 读取属性值
      mapRead['title'] = '1';
      // 修改数据
      Map<String, dynamic> map = Map<String, dynamic>.from(mapRead);
      // 修改数据并赋值
      map['title'] = '2';
    });

    // 支持原始 SQL 语句
    noteDbHelpter.getDatabase().rawUpdate(
        'UPDATE notes SET title = ?, content = ? WHERE _id = ?',
        ['my title', 'my content', 2]).then((int count) {
      print('updated: $count');
    });
  }

}
```

除了常用的增删改查操作，sqflite 同样支持数据库事务操作和批量操作，代码如下。有了对事务操作和批量操作的支持，在批量处理数据时就会方便很多。

```
// 事务操作
await db.transaction((txn) async {
  await txn.execute('CREATE TABLE Test1 (id INTEGER PRIMARY KEY)');
  ...
});

// 批量操作
```

```
batch = db.batch();
batch.insert('Test', {'name': 'item'});
batch.update('Test', {'name': 'new_item'}, where: 'name = ?', whereArgs: ['item']);
batch.delete('Test', where: 'name = ?', whereArgs: ['item']);
results = await batch.commit();
```

sqflite 支持如下数据类型：INTEGER（对应 int）、REAL（对应 num）、TEXT（对应 string）、BLOB（对应 Uint8List）。这几种数据类型的特点与其他编程语言中对应的数据类型特点一致，这里就不再详细介绍了。

关于 Flutter 文件操作、数据库操作、手势操作的内容就介绍这么多，其中比较重点的是文件操作和数据库操作，希望大家多多实践，巩固学习。

第 11 章

Flutter 自定义组件与方法封装

在开发过程中，有些 UI 和需求无法通过现有的 Flutter 组件来实现，这时我们就要自定义组件，实现各种效果。本章将介绍 Flutter 中自定义组件的几种方式，并配合示例详细讲解。同时本章会讲解 Flutter 方法的封装。

11.1 自定义组件

Flutter 支持组件自定义，自定义组件的方式一般有以下几种。

- 通过继承组件修改和扩展组件的功能。
- 通过组合组件扩展组件的功能。
- 使用 CustomPaint 绘制自定义组件。

以上三种方式有各自的优势和特点。相对来说，通过 CustomPaint 绘制自定义组件比较复杂。接下来我们分别介绍以上三种方式。

11.1.1 通过继承组件实现自定义

通过继承组件来实现自定义组件在 Flutter 中非常常见，例如 NetworkImage 组件和 AssetImage 组件都是通过继承 ImageProvider 组件来实现不同功能场景的，Center 组件则是继承自 Align 组件的。我们可以根据实际的使用场景来选择基础组件，在此基础上通过继承实现最终的功能。

这里我们以 Dialog 组件为基础组件，通过继承来实现自定义组件。Dialog 组件继承自

StatelessWidget，是一个无状态基础组件。我们可以在此基础上扩展一个具有全新功能的组件。例如我们想显示具有加载中效果的对话框，代码如下。

```
import 'package:flutter/material.dart';

// 继承 Dialog 组件，具有 Dialog 的特性和方法
class LoadingDialog extends Dialog {
  String text;

  // 建立构造方法，传递参数
  LoadingDialog({Key key, @required this.text}) : super(key: key);

  @override
  Widget build(BuildContext context) {
    // 具体逻辑
    return Material(
      type: MaterialType.transparency,
      child: Center(
        child: SizedBox(
          width: 120.0,
          height: 120.0,
          child: Container(
            decoration: ShapeDecoration(
              color: Color(0xffffffff),
              shape: RoundedRectangleBorder(
                borderRadius: BorderRadius.all(
                  Radius.circular(8.0),
                ),
              ),
            ),
            child: Column(
              mainAxisAlignment: MainAxisAlignment.center,
              crossAxisAlignment: CrossAxisAlignment.center,
              children: <Widget>[
                CircularProgressIndicator(),
                Padding(
                  padding: const EdgeInsets.only(
                    top: 20.0,
                  ),
```

```
                child: Text(
                  text,
                  style: TextStyle(fontSize: 12.0),
                ),
              ),
            ],
          ),
        ),
      ),
    ),
  );
  }
}

// 具体用法
class CustomWidgetSamples extends StatefulWidget {
  @override
  State<StatefulWidget> createState() {
    return CustomWidgetSamplesState();
  }
}

class CustomWidgetSamplesState extends State<CustomWidgetSamples> {
  @override
  void initState() {
    super.initState();
  }

  @override
  Widget build(BuildContext context) {
    return Scaffold(
        appBar: AppBar(title: Text('CustomWidget'), primary: true),
        body: Container(
          child: Align(
            alignment: Alignment.center,
            // 构造组件并传递参数
            child: LoadingDialog(text: '加载中...'),
          ),
        ));
```

```
  }
}

// 只需传递 text 参数
LoadingDialog(text: '加载中...'),
```

运行以上代码，效果如图 11-1 所示。

图 11-1　自定义 Dialog 组件效果

通过以上示例可以看到，利用 Dialog 自身的特点进行定制和扩展就可以形成一个新的组件。通过这样的方式，我们可以实现很多组件化需求，更好地复用组件。

11.1.2　通过组合组件实现自定义

组合组件，顾名思义，就是选择各种 Flutter 的基础组件，进行组合拼装，实现一个可以满足需求的新组件。其实 Flutter 中的很多基础组件也是通过组合组件实现的。

我们来看一个示例，通过组合 Container 组件、Row 组件和 Icon 组件实现一个自定义的 ToolBar 组件——更加灵活的顶部菜单栏组件，代码如下。

```
// 自定义 ToolBar 组件
import 'package:flutter/material.dart';
```

```
class ToolBar extends StatefulWidget implements PreferredSizeWidget {
  // 构造方法，传递参数
  ToolBar({@required this.onTap}) : assert(onTap != null);
  // 属性参数，单击回调
  final GestureTapCallback onTap;

  @override
  State createState() {
    return ToolBarState();
  }
  // AppBar 需要实现 PreferredSizeWidget
  @override
  Size get preferredSize {
    return Size.fromHeight(56.0);
  }
}

class ToolBarState extends State<ToolBar> {
  @override
  Widget build(BuildContext context) {
    // 设置布局
    return SafeArea(
      top: true,
      child: Container(
        color: Colors.blue,
        child: Row(
          children: <Widget>[
            Icon(
              Icons.menu,
              color: Colors.white,
              size: 39,
            ),
            Expanded(
              child: Container(
                color: Colors.white,
                padding: EdgeInsets.all(5),
                margin: EdgeInsets.all(5),
                child: Text(
```

```
                  '搜索...',
                  style: TextStyle(fontSize: 18),
                ),
              ),
            ),
            GestureDetector(
              onTap: this.widget.onTap,
              child: Icon(
                Icons.photo_camera,
                color: Colors.white,
                size: 39,
              ),
            )
          ],
        ),
      ),
    );
  }
}

// 调用 ToolBar
class CustomWidgetSamples extends StatefulWidget {
  @override
  State<StatefulWidget> createState() {
    return CustomWidgetSamplesState();
  }
}

class CustomWidgetSamplesState extends State<CustomWidgetSamples> {
  @override
  void initState() {
    super.initState();
  }

  @override
  Widget build(BuildContext context) {
    return Scaffold(
        // 自定义 ToolBar
        appBar:ToolBar(
```

```
          onTap: () {
            print('click');
          },
        ),
        primary: true,
        body: Column(
          children: <Widget>[
            Container(
              child: Align(
                alignment: Alignment.center,
                child: LoadingDialog(text: '加载中...'),
              ),
            )
          ],
        ));
  }
}
```

运行以上代码，效果如图 11-2 所示。

图 11-2　自定义 ToolBar 组件效果

11.1.3　通过 CustomPaint 绘制组件

本节将介绍通过 CustomPaint 来绘制自定义组件的方式。这种方式相对于前两种方式来说比较复杂，适用于实现复杂效果的场景，例如绘制不规则的图形、布局、图表等。

CustomPaint 继承自 SingleChildRenderObjectWidget，其构造方法如下。

```
const CustomPaint({
    Key key,
    // 背景画笔
    this.painter,
    // 前景画笔
    this.foregroundPainter,
    // 画布尺寸
    this.size = Size.zero,
    // 判断是否复杂，Flutter 会设置一些缓存优化策略
    this.isComplex = false,
    // 下一帧是否改变
    this.willChange = false,
    // 子组件，可以为空
    Widget child,
  })
```

上述构造方法中比较重要的是背景画笔和前景画笔，我们通过这两个画笔可以实现自定义绘制。

通过 CustomPaint 绘制组件的核心是自定义 CustomPainter，CustomPainter 是画笔工具，用于在画布上绘制内容。我们要了解 CustomPainter 里面的方法结构，通过继承 CustomPainter 并重写里面的核心方法即可绘制组件。继承、重写 CustomPainter 的核心代码如下。

```
class Sky extends CustomPainter {
  // 绘制方法
  @override
  void paint(Canvas canvas, Size size) {
    // canvas 为画布
    // size 为画布大小
  }

  // 判断刷新布局时是否重绘，根据实际情况返回值
  @override
```

```
  bool shouldRepaint(Sky oldDelegate) => false;
}
```

以上代码中最重要的就是通过画布渲染展现绘制效果。Flutter 中的 Canvas 对象和其他平台的 Canvas 对象功能、作用基本一样，其中包含很多绘制方法，这里大致列出一些。

```
// 保存画布内容
void save() native 'Canvas_save';
void saveLayer(Rect bounds, Paint paint);
void restore() native 'Canvas_restore';
int getSaveCount() native 'Canvas_getSaveCount';
// 位移
void translate(double dx, double dy) native 'Canvas_translate';
// 缩放
void scale(double sx, [double sy]) => _scale(sx, sy ?? sx);
// 旋转
void rotate(double radians) native 'Canvas_rotate';
void skew(double sx, double sy) native 'Canvas_skew';
void transform(Float64List matrix4);
// 裁剪
void clipRRect(RRect rrect, {bool doAntiAlias = true});
void clipPath(Path path, {bool doAntiAlias = true});
// 绘制
void drawColor(Color color, BlendMode blendMode);
void drawLine(Offset p1, Offset p2, Paint paint);
void drawPaint(Paint paint);
void drawRect(Rect rect, Paint paint);
void drawOval(Rect rect, Paint paint);
void drawCircle(Offset c, double radius, Paint paint);
void drawArc(Rect rect, double startAngle, double sweepAngle, bool useCenter,
Paint paint);
void drawPath(Path path, Paint paint);
void drawImage(Image image, Offset p, Paint paint);
void drawShadow(Path path, Color color, double elevation, bool transparentOccluder);
```

在进行画布绘制时，需要使用画笔 Paint 组件，我们可以创建相应的画笔在画布上进行绘制。Paint 组件也有很多可以设置的属性，常用的有以下几种。

- color：画笔颜色。
- style：绘制模式，画线或充满。

- maskFilter：绘制完成但没有被混合到布局时添加的遮罩效果，如 blur 效果。
- strokeWidth：画笔宽度。
- strokeCap：画笔笔触类型。
- shader：着色器，一般用来绘制渐变效果。

接下来我们来看一下如何配置 Paint 组件属性，以及如何使用 Paint 组件，代码如下。

```
Paint myPaint = Paint()
    ..color = Colors.blueAccent // 画笔颜色
    ..strokeCap = StrokeCap.round // 画笔笔触类型
    ..isAntiAlias = true // 是否启动抗锯齿
    ..blendMode = BlendMode.exclusion // 颜色混合模式
    ..style = PaintingStyle.fill // 绘制模式，默认为填充
    ..colorFilter = ColorFilter.mode(Colors.blueAccent,
        BlendMode.exclusion) // 颜色渲染混合模式
    ..maskFilter = MaskFilter.blur(BlurStyle.inner, 2.0) // 模糊遮罩效果
    ..filterQuality = FilterQuality.high // 颜色渲染模式的质量
    ..strokeWidth = 10.0; // 画笔宽度
```

了解了 Paint 组件的属性配置后，通过 CustomPaint 来自定义组件就比较容易了，接下来我们来绘制一个简单的圆角矩形组件，代码如下。

```
class CustomWidgetSamplesState extends State<CustomWidgetSamples> {
  @override
  void initState() {
    super.initState();
  }

  @override
  Widget build(BuildContext context) {
    return Scaffold(
        appBar: AppBar(
          title: Text('CustomWidget'),
        ),
        body: Column(
          children: <Widget>[
            CustomPaint(
              painter: Sky(),
              child: Center(
```

```
                child: Text(
                  '文字',
                ),
              ),
            )
          ],
        ));
  }
}

class Sky extends CustomPainter {
  @override
  void paint(Canvas canvas, Size size) {
    // 绘制圆角矩形
    // 用 Rect 构建一个边长 60，中心点坐标为(150,150)的矩形
    Rect rectCircle =
        Rect.fromCircle(center: Offset(150.0, 150.0), radius: 60.0);
    // 根据上面的矩形构建一个圆角矩形
    RRect rrect = RRect.fromRectAndRadius(rectCircle, Radius.circular(30.0));
    canvas.drawRRect(rrect, Paint()..color = Colors.yellow);
  }

  @override
  bool shouldRepaint(Sky oldDelegate) => false;
  @override
  bool shouldRebuildSemantics(Sky oldDelegate) => false;
}
```

运行以上代码，效果如图 11-3 所示。

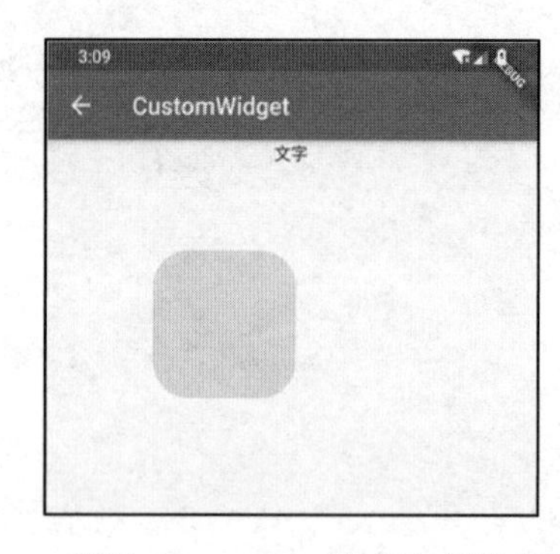

图 11-3　通过 CustomPaint 绘制的圆角矩形组件

11.2　方法封装

同很多编程语言一样，Flutter 也支持一些工具类、方法的封装，方便我们进行使用，使代码逻辑更加清晰。Flutter 方法的封装和 Java 有点像，我们可以把一些常用的方法封装成一个工具类，使用时直接实例化调用即可，这也体现了面向对象编程的封装特性。

这里我们直接创建一个工具类来展示 Flutter 方法的封装，代码如下。

```
import 'package:flutter/services.dart';
import 'package:flutter/widgets.dart';

class Utils {
  BuildContext context;

  // 设置构造方法，传递参数，通过 key:value 形式
  Utils({@required this.context}) : assert(context != null);

  // 指定返回类型，定义方法名
  /// 获取时间戳毫秒数，13 位
  int getMilliseconds() {
    return DateTime.now().millisecondsSinceEpoch;
  }
  // 方法名后可以设置要传递的参数
  /// 复制到剪贴板
  void setClipData(String text) {
    Clipboard.setData(ClipboardData(text: text));
  }

  // 以下画线开始的方法名只能在类内部调用
  /// 获取屏幕宽度
  double _getScreenWidth(BuildContext context) {
    return MediaQuery.of(context).size.width;
  }

  /// 获取屏幕高度
  double getScreenHeight(BuildContext context) {
    return MediaQuery.of(context).size.height;
  }
```

```
  /// 获取屏幕状态栏高度
  double getStatusBarTop(BuildContext context) {
    return MediaQuery.of(context).padding.top;
  }

  /// 获取屏幕方向
  Orientation getScreenOrientation(BuildContext context) {
    return MediaQuery.of(context).orientation;
  }

  Future<String> getBatteryLevel() async {
    var batteryLevel = 'unknown';
    MethodChannel methodChannel = MethodChannel('samples.flutter.io/battery');
    try {
      int result = await methodChannel.invokeMethod('getBatteryLevel');
      batteryLevel = 'Battery level: $result%';
    } on PlatformException {
      batteryLevel = 'Failed to get battery level.';
    }
    return batteryLevel;
  }
}

// 使用时，构造方法传递参数通过 key：value 形式完成
Utils utils = Utils(context: context);

// 调用方法
utils.getScreenHeight(context);
```

通过以上代码就实现了对 Flutter 基本方法的封装。利用 Flutter 的封装特性，我们不仅可以封装工具类，还可以封装网络请求、组件、页面等。在学习 Flutter 的封装特性时，我们可以参照其他面向对象语言的封装特性，对比学习。

第12章

Flutter 动画的实现

如果想让应用或产品的用户体验变得更好，动画效果是一个很重要的因素。Flutter 也提供了对大部分动画效果的支持，如基础的渐变动画、位移动画、旋转动画等，以及特有的 Hero 动画。Hero 动画的作用是，在页面跳转切换时，使某个组件具有过渡跳转效果，因此 Hero 动画也称共享元素过渡动画。

本章将介绍 Flutter 中动画的基本使用方法和特点，基础动画和 Hero 动画都会涉及，同时会配合一些示例进行讲解。

12.1 动画简介

动画这个概念在大部分编程语言里都存在，也都有对应的实现方式。一般的动画都会有平移、缩放、旋转、透明度渐变等不同效果，在这些基础效果上进行扩充还能实现更加丰富的动画效果。

动画是由动画帧构成的，帧率（每秒帧数）FPS 对于动画流畅度起着很重要的作用，一般帧率越高动画越流畅。从人眼观看角度来说，帧率超过 16FPS，动画就比较流畅了。帧率超过 32FPS，动画非常流畅、细腻。

Flutter 中的动画效果一般通过 Animation、AnimationController、CurvedAnimation、Tween 等组件实现，接下来我们逐一介绍。

1. Animation

Animation 是一个泛型类型，也是一个抽象类，可以写成 Animation<int>、Animation<double>、

Animation<Offset>、Animation<Color>、Animation<Size> 等形式，用于生成指导动画。Animation 支持对动画的操作方法、状态等进行操作，主要功能是保存动画的插值和状态。在动画的每一帧中，我们可以通过 Animation 的 value 属性获取动画的当前状态值。

我们还可以在 Animation 上添加对动画的监听，具体如下。

- addListener()：用于监听动画每一帧，动画每一帧发生变化时都会回调这个监听方法。
- addStatusListener()：用于监听动画状态的改变，如开始、结束、正向运动、反向运动等。

2. AnimationController

顾名思义，AnimationController 是动画生成控制类，是一个特殊的 Animation 组件，继承自 Animation<double>。AnimationController 主要负责动画的控制，如开始、停止、反向播放等，也可以控制屏幕使其每刷新一帧便生成一个新的数值。数值的生成与屏幕刷新有关，每秒通常会产生 60 个数值，默认范围是 0.0~1.0。

AnimationController 中的几个常用方法如下。

- forward()：用于控制动画的开始。
- stop()：用于控制动画的停止。
- reverse()：用于控制动画的反向播放。

由于 AnimationController 继承自 Animation<double>，所以它也具有 Animation 的方法和作用。我们来看一个 AnimationController 的简单用法示例，代码如下。

```
// 动画控制类，指定数值的上下范围，这里是在 3 秒时间内产生 0~2 的小数
final AnimationController _controller = AnimationController(
      lowerBound: 0,
      upperBound: 2,
      duration: const Duration(seconds: 3),
      vsync: this);

// 不指定范围，默认产生 0.0~1.0 的数字，Duration 时间参数设置为 3 秒
final AnimationController _valueController =
      AnimationController(duration: const Duration(seconds: 3), vsync: this);
```

以上代码中有 vsync: this 属性，this 指的是当前的类对象，所以类中需要实现相关的方法，

在本例中我们要在定义类时实现 TickerProviderStateMixin 方法或 SingleTickerProviderStateMixin 方法，代码如下。

```
class AnimationSamplesState extends State<AnimationSamples>
    with TickerProviderStateMixin {
    // 或者 with SingleTickerProviderStateMixin
```

在创建 AnimationController 时要传入 vsync 参数，vsync 参数的作用是防止执行不必要的动画操作，消耗不必要的资源。例如，锁屏时一般无须设置动画。

3. CurvedAnimation

CurvedAnimation 也继承自 Animation<double>，主要作用是对 AnimationController 产生的数值进行不同的曲线变化操作，也就是将动画的运动轨迹变成一个曲线，实现各式各样的运动效果。CurvedAnimation 的大致用法如下。

```
// 需要配合 AnimationController 使用
final Animation<double> animation = CurvedAnimation(
  parent: controller,
  // 数值产生的曲线效果
  curve: Curves.easeIn,
  // 反向数值产生的曲线效果（动画倒放）
  reverseCurve: Curves.easeOut,
);
```

数值产生的曲线效果如图 12-1 所示。

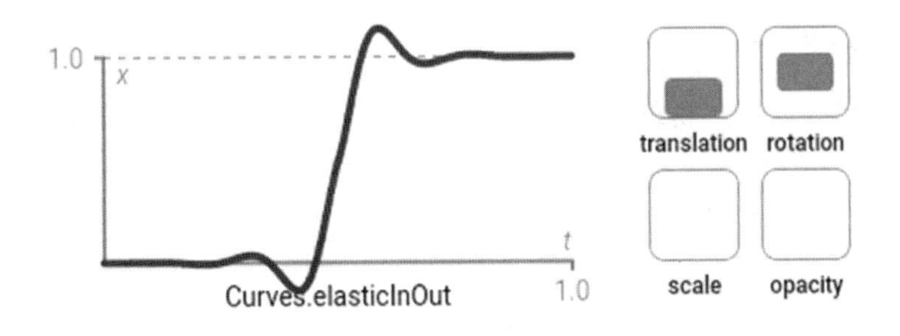

图 12-1　数值产生的曲线效果

除了以上效果，CurvedAnimation 还提供了很多其他曲线效果，常用的如下。

```
class Curves {
  // 匀速
```

```
  static const Curve linear = _Linear._();
  // 匀减速
  static const Curve decelerate = _DecelerateCurve._();
  // 先快速匀速，然后缓慢变成 EaseIn 曲线效果
  static const Cubic fastLinearToSlowEaseIn = Cubic(0.18, 1.0, 0.04, 1.0);
  // 先加速，后减速
  static const Cubic ease = Cubic(0.25, 0.1, 0.25, 1.0);
  // 先慢速，后快速
  static const Cubic easeIn = Cubic(0.42, 0.0, 1.0, 1.0);
  // 先慢速，后快速，最后匀速
  static const Cubic easeInToLinear = Cubic(0.67, 0.03, 0.65, 0.09);
}
```

除了官方自带的曲线效果，我们也可以自定义数值产生的曲线效果，只要继承 Curve 对象并实现里面的 transform 方法即可，我们来看一个简单的震动曲线效果的实现，代码如下。

```
class ShakeCurve extends Curve {
  @override
  double transform(double t) {
    return math.sin(t * math.PI * 3);
  }
}
```

4. Tween

默认情况下，AnimationController 产生的数值范围是 0.0~1.0。如果我们想生成其他范围的数值该怎么办呢？这时，Tween 就可以派上用场了。这里需要注意，Tween 继承自 Animatable<T>，而不继承自 Animation<T>。

通过 Tween，我们可以实现很多渐变效果。例如，可以使用 Tween 生成–100~100 的数值、从白色到蓝色的颜色数值等。Tween 中有很多封装子类，如 IntTween、ColorTween、AlignmentGeometryTween、DecorationTween、TextStyleTween、RelativeRectTween、RectTween 等。借助这些子类，我们就可以实现非常丰富的渐变动画效果。

先来看一下 Tween 的简单用法，代码如下。

```
// 声明定义 Tween
final Tween doubleTween = Tween<double>(begin: -100.0, end: 100.0);
final Tween intTween = IntTween(begin: 0, end: 255);
final Tween colorTween = ColorTween(begin: Colors.orange, end: Colors.teal);
```

```
// 调用，需要用到 AnimationController
final AnimationController controller = new AnimationController(
duration: const Duration(milliseconds: 500), vsync: this);

// 直接调用 animate，绑定到 AnimationController
colorTween .animate(controller);

// Tween 还可以配合曲线效果 CurvedAnimation
Animation<double> _doubleAnimation = Tween<double>(begin: -100.0, end: 100.0).animate
(
      CurvedAnimation(
        parent: _valueController,
        // 生成数据对应的速率曲线
        curve: Curves.easeIn,
      ),
    )..addListener(() {
        setState(() {
          // 监听获取值并更新 UI
        });
      });
    })..addStatusListener((AnimationStatus status){
          // 状态值有 dismissed、forward、reverse、completed 几种
    });
```

通过以上代码，我们就实现了具有颜色渐变效果的简单动画。

5. 其他常用组件

当我们想要实现不同的动画效果时，如果每次都要重复编写逻辑可能很麻烦，所以 Flutter 封装好了一些常用的动画效果组件，如 AnimatedBuilder、AnimatedModalBarrier、FadeTransition、DecoratedBoxTransition、PositionedTransition、RelativePositionedTransition、RotationTransition、ScaleTransition、SizeTransition、SlideTransition 等，这些动画效果组件都继承自 AnimatedWidget，而 AnimatedWidget 继承自 StatefulWidget，是有状态组件。

按照这个原理，我们可以自定义实现具有某种效果的 AnimatedWidget。例如，我们可以实现一个具有放大效果的组件，编写里面的核心方法，代码如下。

```
// 自定义实现一个具有放大效果的组件
class ScaleAnimatedWidget extends AnimatedWidget {
  // 需要定义一个 AnimationController
```

```
  final AnimationController animationController;
  // 构造方法
  ScaleAnimatedWidget(
      {Key key,
      Animation<double> animation,
      @required this.animationController})
      : super(key: key, listenable: animation);

  @override
  createState() {
    animationController.forward();
    return super.createState();
  }
  // 核心动画效果逻辑
  @override
  Widget build(BuildContext context) {
    final Animation<double> animation = listenable;
    return Center(
      child: Container(
        decoration: BoxDecoration(color: Colors.redAccent),
        margin: EdgeInsets.symmetric(vertical: 10.0),
        height: animation.value * 100,
        width: animation.value * 100,
      ),
    );
  }
}
```

我们利用 AnimatedWidget 实现了具有放大效果的组件，可以发现，代码中只传入 animation 方法和 animationController 方法，然后构建一个动画组件，还是不够灵活，不具有通用性。如果能把一个组件传进来并给它设置动画，这样会更具有通用性和灵活性。此时可以使用 AnimatedBuilder 进行动画组件封装，以实现上述需求。

Flutter 中使用 AnimatedBuilder 封装的组件有 BottomSheet、ExpansionTile、PopupMenu、ProgressIndicator、RefreshIndicator、Scaffold、SnackBar、TabBar、TextField 等。

AnimatedBuilder 的灵活性和通用性更高，下面我们来实现一个可以传入自定义组件的动画组件，具体代码如下。

```
class GrowTranstion extends StatelessWidget {
```

```
  final Widget child;
  final Animation<double> animation;

  GrowTranstion(this.animation, this.child);
   // 核心逻辑
  @override
  Widget build(BuildContext context) {
    return Center(
      child: AnimatedBuilder(
        animation: animation,
        builder: (BuildContext context, Widget child) {
          return Container(
            child: child,
          );
        },
        child: child,
      ),
    );
  }
}

// 使用时直接传入 child 组件，具有通用性
Widget build(BuildContext context) {
    return GrowTransition(child: LogoWidget(), animation: animation);
  }
...
// 还可传入其他类型组件，比较灵活
Widget build(BuildContext context) {
    return GrowTransition(child: ImageWidget(), animation: animation);
  }
```

12.2　基础动画

上一节介绍了关于 Flutter 动画的基础内容，本节将通过一些示例来巩固 Flutter 基础动画的实现方法。首先来看一下组件动态缩放效果的实现代码，如下。

```
  // 定义 AnimationController
  AnimationController _valueController =
        AnimationController(duration: const Duration(seconds: 3), vsync: this);
```

```
// 根据需要可以定义一个 Animation
Animation<double> _doubleAnimation = Tween<double>(begin: 0.0, end: 100.0).animate(
    CurvedAnimation(
      parent: _valueController,
      // 生成数据速率曲线
      curve: Curves.easeIn,
    ),
  );

 // 对 doubleAnimation 设置动画状态监听
  _doubleAnimation.addStatusListener((AnimationStatus status) {
    if (status == AnimationStatus.forward) {
      print('Animation Start');
    } else if (status == AnimationStatus.completed) {
      print('Animation Completed');
      // _controller.reverse();
    } else if (status == AnimationStatus.reverse) {
      print('Animation Reverse');
    } else if (status == AnimationStatus.dismissed) {
      print('Animation Dismissed');
      _controller.forward();
    }
  });

  // 调用开始动画方法
  _valueController.forward();
...

@override
Widget build(BuildContext context) {
  return Scaffold(
    appBar: AppBar(
      title: Text('Animation'),
    ),
    body: animation(context),
  );
}
```

```
Widget animation(BuildContext context) {
  return Container(
    // 动态设置宽高，实现组件动态缩放效果
    width: _doubleAnimation.value,
    height: _doubleAnimation.value,
    color: Colors.teal,
    margin: EdgeInsets.all(10),
    child: FlutterLogo(),
  );
}
```

通过以上代码，我们就实现了一个简单的组件动态缩放效果，即缩放动画。同样地，我们可以按照这个方式来实现旋转、渐变、位移、颜色变化等复杂动画。对于以上功能，我们还可以使用 Flutter 封装好的缩放动画组件 ScaleTransition 来实现。ScaleTransition 继承自 AnimatedWidget，其用法如下。

```
// 定义一个 AnimationController
AnimationController _valueController =
      AnimationController(duration: const Duration(seconds: 3), vsync: this);

Animation<double> _scaleAnimation = Tween(begin: 0.0, end: 2.0).animate(_valueContr
oller);

  // 开始动画
  _valueController.forward();

Widget scaleAnimation(BuildContext context) {
  // 直接使用 ScaleTransition 组件
  return ScaleTransition(
    scale: _scaleAnimation,
    child: Text('ScaleTransition'),
  );
}
```

通过对比以上两段代码可以看出，使用 Flutter 已经封装好的动画组件更加简单，可扩展性更强。旋转、渐变、位移类动画组件的使用方法类似，这里就不重复讲解了。希望感兴趣的读者可以亲自动手尝试，多多实践。

12.3　Hero 动画

Hero 动画也称共享元素过渡动画，主要用于实现页面跳转时某个组件的过渡跳转效果。

用户从页面中选择一个元素（通常是一个图像），然后打开所选元素的详情页面，这个过程中对元素和页面设置的动画就是 Hero 动画。举例来说，某个页面上有一个用户头像，单击头像会跳转到另一个页面，此时可以在用户头像上设置一个动画，在新页面打开时也设置一个过渡动画，如让头像渐渐移动到底部消失，让目标页面渐渐显示。

我们先来看一下 Hero 动画的基本使用方式，如下。

- 在页面 A 和页面 B 上分别定义 Hero 组件，设置相同的 tag 值进行匹配。
- 在路由里进行配置，从页面 A 跳转到页面 B。
- 单击跳转，执行动画。

要掌握 Hero 动画的实现方法，我们需要先了解 Hero 动画的执行过程：根据两个 Hero 组件计算出一个补间矩形，将这个补间矩形作为中间遮罩层实现过渡动画。在跳转过程中，页面 A 的 Hero 组件会跳转到中间遮罩层，然后进入页面 B。

下面我们用 Flutter 官方 Hero 动画示意图来深入理解执行过程。

在跳转前状态下，路由中只有一个源路由（Source route）页面和一个空的补间矩形（Empty），这个补间矩形作为中间遮罩层（Overlay），目标路由（Destination route）页面中还没有 Hero 动画（Doesn't exist yet），如图 12-2 所示。

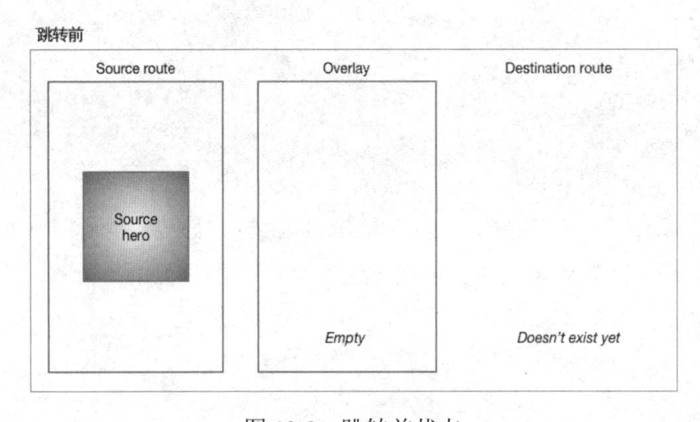

图 12-2　跳转前状态

Hero 动画刚开始执行时，源路由页面的源 Hero 动画（Source hero）逐渐开始运动，中间遮罩层也渲染了目标路由页面的 Hero 动画，此时目标路由页面内的 Hero 动画还没开始展示，如图 12-3 所示。

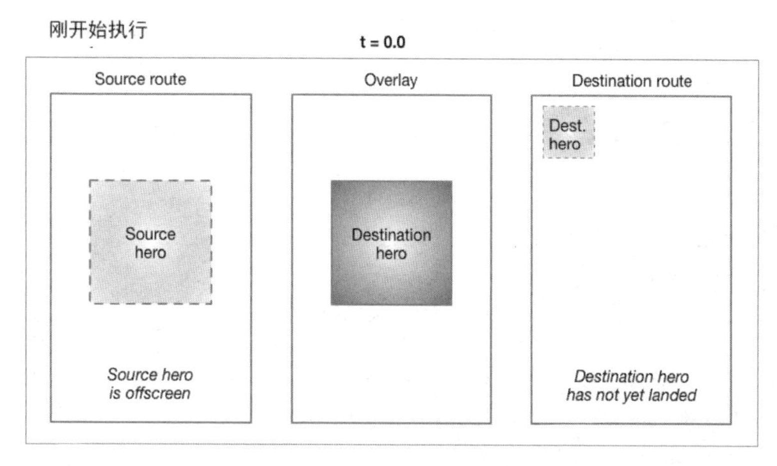

图 12-3　Hero 动画刚开始执行

在 Hero 动画执行过程中，动画运动一段时间后，源路由页面的 Hero 动画渐渐移出屏幕，中间遮罩层内目标路由页面的 Hero 动画逐渐进入屏幕渲染，目标路由页面内的 Hero 动画刚刚准备进入，还未完全展示，如图 12-4 所示。

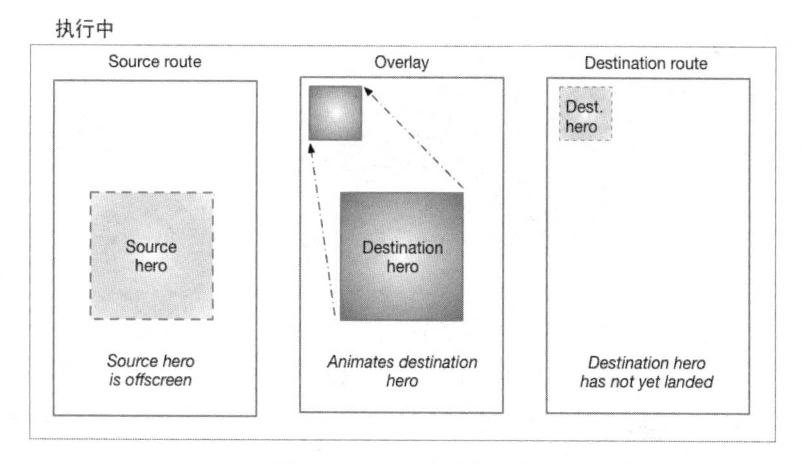

图 12-4　Hero 动画执行中

最后，Hero 动画执行完毕，页面跳转结束，中间遮罩层动画执行完毕，成为空层，目标路由页面的 Hero 动画完全显示出来，如图 12-5 所示。

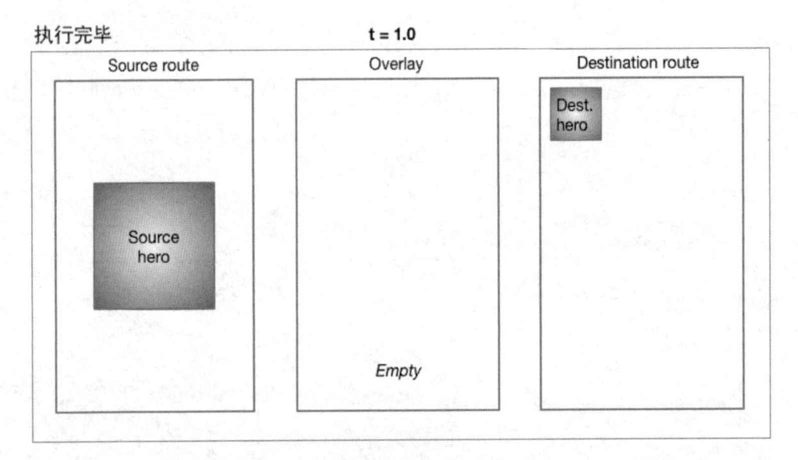

图 12-5　Hero 动画执行完毕

通过上面的 Hero 动画执行过程示意图，相信大家对 Hero 动画的实现方式已经有了一定的了解。接下来我们介绍 Hero 动画的构造方法，如下。

```
const Hero({
    Key key,
    // 标识标签
    @required this.tag,
    // 转变动画
    this.createRectTween,
    // 飞行过程中的组件，可以自定义
    this.flightShuttleBuilder,
    // 构造占位组件
    this.placeholderBuilder,
    // 手势滑动返回时是否有 Hero 动画
    this.transitionOnUserGestures = false,
    @required this.child,
  })
```

可以看到，Hero 动画的构造方法并不复杂，使用时非常简单：源路由页面和目标路由页面都用 Hero 组件来包裹一个组件，然后设置相同的 tag 值使两个页面匹配，以下是一个具体示例。

```
// 在页面 A 定义一个 Hero 组件
// 如果内部组件用的是 InkWell，则 InkWell 外面要用 Material 组件包裹
...
Hero(
      // 相同 tag 值
      tag: "iconTag",
```

```
      child: Material(
        child: InkWell(
          child: Icon(
            Icons.room,
            size: 70.0,
          ),
          // 单击
          onTap: () {
            gotoPage();
          },
        ),
      ));
...
// 单击后进行路由跳转
  void gotoPage() {
    Navigator.push(context, MaterialPageRoute(builder: (context) {
      return HeroSamples();
    }));
  }
...

// 页面 B，定义一个具有相同 tag 值的 Hero 组件
class HeroSamplesState extends State<HeroSamples> {
  @override
  void initState() {
    super.initState();
  }

  @override
  Widget build(BuildContext context) {
    return Scaffold(
      appBar: AppBar(title: Text('Hero'), primary: true),
      body: Column(
        children: <Widget>[
          Hero(
            // 要设置相同的 tag 值
            tag: "iconTag",
            child: Icon(
              Icons.room,
```

```
                size: 70.0,
              ),
            ),
          ],
        ),
      );
    }
  }
```

在 Hero 动画的基础上，我们来拓展讲解 Hero 径向动画，也就是从圆形组件经过动画变换成为矩形组件的过程，其原理如图 12-6 所示。

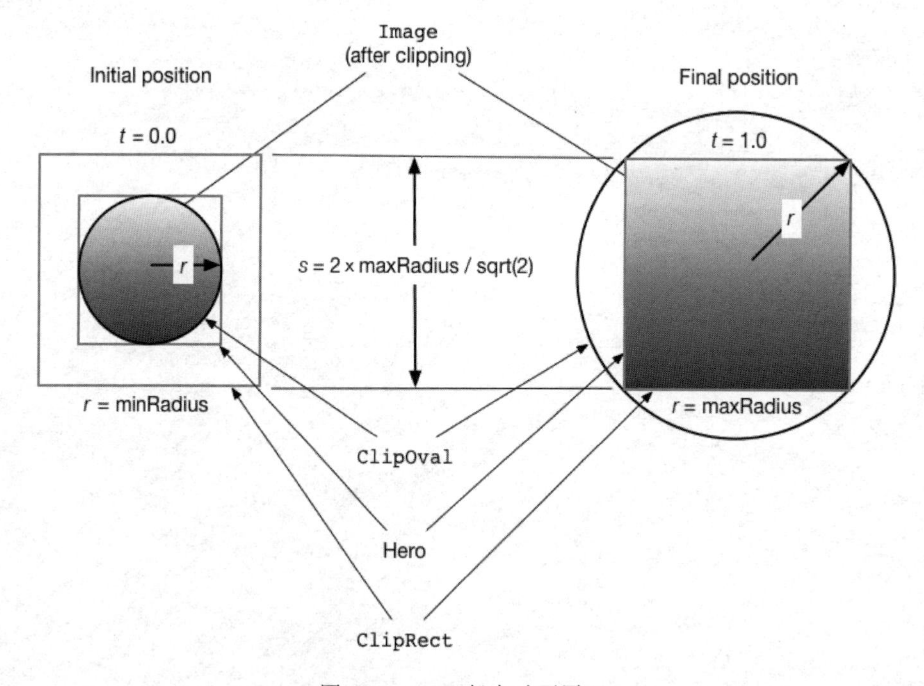

图 12-6　Hero 径向动画原理

从图 12-6 中可以看出，Hero 径向动画的执行过程是，将圆形组件的最小半径（minRadius）扩大到目标矩形组件所在外切圆的最大半径（maxRadius），然后再以这个最大半径作为矩形对角线的一半，裁剪成矩形目标组件。

Hero 径向动画效果的简单实现方法如下。

```
Hero(
    tag: "radialTag",
```

```
      child: Material(
        child: InkWell(
          child: ClipOval(
            child: SizedBox(
              width: 100,
              height: 100,
              child: ClipRect(
                child: Image.asset(
                  "assets/image_appbar.jpg",
                  fit: BoxFit.contain,
                ),
              ),
            ),
          ),
          onTap: () {
            gotoPage();
          },
        ),
      ))

...

// 目标路由页面
Hero(
  tag: "radialTag",
  child: Image.asset(
    "assets/image_appbar.jpg",
    fit: BoxFit.contain,
  ),
)
```

12.4　交错动画

Flutter 官方推出了交错动画（StaggeredAnimation），即将很多个不同的动画效果叠加在一起同时控制，也可以理解为一系列动画的组合，其原理如图 12-7 所示。

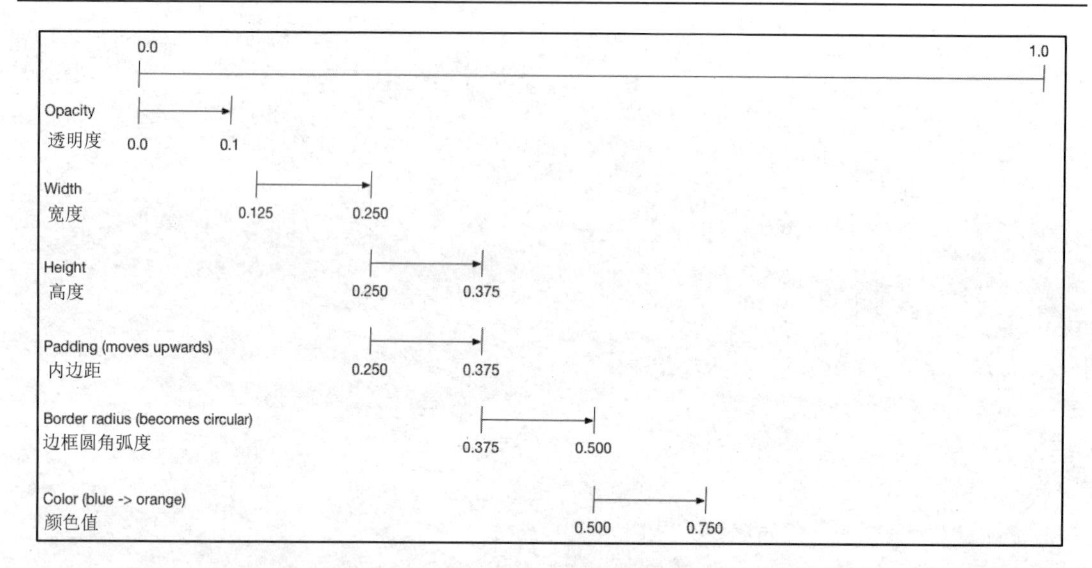

图 12-7　交错动画原理

通过图 12-7 大致可以看出，交错动画是一系列不同动画效果的组合，除了基础的动画组合，还可以让各个动画在时间上进行叠加。实现交错动画效果需要创建多个动画对象，用一个 AnimationController 组件控制所有动画。以下是一个官方示例，将多种简单动画组合成一个复杂的交错动画，具体代码如下。

```
import 'package:flutter/material.dart';

class StaggerAnimationSamples extends StatefulWidget {
  @override
  State<StatefulWidget> createState() {
    return StaggerAnimationSamplesState();
  }
}

class StaggerAnimationSamplesState extends State<StaggerAnimationSamples>
with TickerProviderStateMixin {
  // 需要创建一个 AnimationController 组件
  AnimationController _controller;
  // 根据需要创建不同效果的 Animation 对象
  Animation<double> opacity;
  Animation<double> width;
  Animation<Color> color;
```

```
  @override
  void initState() {
super.initState();
// 实例化 AnimationController 组件
    _controller = AnimationController(
        duration: const Duration(milliseconds: 5000), vsync: this);

    // 实例化动画效果，组合起来
    opacity = Tween<double>(
      begin: 0.0,
      end: 1.0,
    ).animate(
      CurvedAnimation(
        parent: _controller,
        curve: Curves.ease,
      ),
    )..addListener(() {
        setState(() {
          // 监听获取值并更新 UI
        });
      });

    width = Tween<double>(
      begin: 50.0,
      end: 150.0,
    ).animate(
      CurvedAnimation(
        parent: _controller,
        curve: Curves.linear,
      ),
    )..addListener(() {
        setState(() {
          // 监听获取值并更新 UI
        });
      });
    color = ColorTween(
      begin: Colors.blue,
      end: Colors.teal,
    ).animate(
```

```
    CurvedAnimation(
      parent: _controller,
      curve: Interval(
        0.0,
        0.8,
        curve: Curves.ease,
      ),
    ),
  )..addListener(() {
      setState(() {
        // 监听获取值并更新 UI
      });
    });
}

// 控制动画的播放
Future<Null> _playAnimation() async {
  try {
    await _controller.forward().orCancel;
    await _controller.reverse().orCancel;
  } on TickerCanceled {
    // 动画取消或停止时执行
  }
}

@override
Widget build(BuildContext context) {
  return Scaffold(
    appBar: AppBar(title: Text('StaggerAnimation'), primary: true),
    body: Column(
      children: <Widget>[
        RaisedButton(
          child: Text("开始"),
          onPressed: () {
              // 单击播放动画
            _playAnimation();
          },
        ),
        // 页面组件可借助 Animation 效果值进行变换
```

```
          Container(
            width: 300.0,
            height: 300.0,
            child: Container(
              alignment: Alignment.bottomCenter,
              child: Opacity(
                opacity: opacity.value,
                child: Container(
                  width: width.value,
                  height: width.value,
                  decoration: BoxDecoration(
                    color: color.value,
                    border: Border.all(
                      color: Colors.indigo[300],
                      width: 3.0,
                    )
                  ),
                ),
              ),
            ),
          ),
        ],
      ),
    );
  }
}
```

在以上代码中，各种动画效果被组合并被绑定到了同一个 AnimationController 组件上，实现了通过一个 AnimationController 组件来控制多个动画变化，最终实现一个交错动画的效果。这段代码中用到了很多前面已介绍的基础知识，其实并不难，希望大家仔细研究并实践巩固。

第 13 章

Flutter 主题与应用国际化

主题这个概念相信大家都不陌生，我们在使用移动应用或网页时，都有夜间模式主题、护眼模式主题等。在 Flutter 中，主题的概念和作用也是一样的。我们可以设置全局或局部主题样式、字体样式，使得应用在整体上具有一定的样式风格。

国际化可以让应用支持多种语言，例如运行在国内时使用中文简体，运行在国外时使用英文、日文等。我们也可以将国际化处理称为本地化处理。Flutter 原生 API 支持国际化处理，也可以用官方提供的插件库来实现。

13.1 主题的实现

要想实现 Flutter 全局主题，只要在应用入口处创建一个全局主题配置项即可。当然 Flutter 也支持局部主题的自定义配置，可为应用的某几个页面设置单独的个性化主题。我们可以通过主题 API 从应用的整体色调、夜间模式、护眼模式、字体样式等方面进行个性化主题设置。本节将介绍创建全局主题、设置局部主题、扩展和修改全局主题的方法。

13.1.1 创建全局主题

Flutter 支持为应用创建全局主题，实现时需要配合 MaterialApp 的 theme 属性。主题在 Flutter 中主要是通过 ThemeData 组件来实现的。

先来看一个创建全局主题的简单示例，代码如下。

```
/// 创建全局主题
Widget theme1() {
```

```
  return MaterialApp(
    /// ThemeData.dark()
    /// ThemeData.light()
    /// 默认有以上两种主题
    /// 也可以自定义主题
theme: ThemeData(
 // 主题色调
      brightness: Brightness.light,
      primaryColor: Colors.lightBlue[800],
      accentColor: Colors.cyan[600],
      backgroundColor: Colors.white70,
    ),
    home: Scaffold(
      appBar: AppBar(
        title: Text('Theme'),
      ),
      body: Text(
        'data',
        style: TextStyle(fontSize: 18, decoration: TextDecoration.none),
      ),
    ),
  );
}
```

ThemeData 提供了 ThemeData.dark()和 ThemeData.light()两种已经配置好的主题色调，可以直接使用。一般情况下，我们可以配置应用的主题色调、主体色、组件前景色调、主体背景色。如果在 MaterialApp 里没有设置全局主题，Flutter 会提供一个默认的全局主题。

ThemeData 包含的可配置属性参数非常多，以下列举常用属性参数。在实际应用时，可以按照需求选择不同属性配置个性化主题。

```
factory ThemeData({
    // 确定是深色调还是浅色调
    Brightness brightness,
    // 主体颜色样本，可用于导航栏的背景色等
    MaterialColor primarySwatch,
    // 主体背景色
    Color primaryColor,
    // 背景色的亮度
    Brightness primaryColorBrightness,
```

```
    // 背景色的亮色版本
    Color primaryColorLight,
    // 背景色的暗色版本
    Color primaryColorDark,
    // 前景色（文本、按钮等）
    Color accentColor,
    // Scaffold 的默认背景色
    Color scaffoldBackgroundColor,
    // BottomAppBar 的默认颜色
    Color bottomAppBarColor,
    // 分割线颜色
    Color dividerColor,
    // 选中高亮的颜色
    Color highlightColor,
    // 选中波纹颜色
    Color splashColor,
    // 非活动/未选中时的颜色
    Color unselectedWidgetColor,
    // 不可用时颜色
    Color disabledColor,
    // 文字选中时颜色
    Color textSelectionColor,
    // 指针光标颜色
    Color cursorColor,
    // 字体
    String fontFamily,
    // 文本样式主题
    TextTheme textTheme,
    TextTheme primaryTextTheme,
    TextTheme accentTextTheme,
    // AppBar 主题
    AppBarTheme appBarTheme,
    // BottomAppBar 主题
    BottomAppBarTheme bottomAppBarTheme,
    ColorScheme colorScheme,
    // Dialog 主题
DialogTheme dialogTheme,
… …
 })
```

接下来，我们通过 ThemeData 来简单配置一个个性化的全局主题，代码如下。

```
import 'package:flutter/material.dart';

void main() => runApp(MyApp());

class MyApp extends StatelessWidget {

  // 构建页面
  @override
  Widget build(BuildContext context) {
    // 入口页面使用 MaterialApp 脚手架实现
    // MaterialApp 脚手架默认自带顶部 ToolBar、路由、主题、国际化等配置
    return MaterialApp(
      title: 'Flutter Demo',
      theme: ThemeData(
        // 配置并应用全局主题
        brightness: Brightness.light,
        // 通过 primarySwatch 属性配置状态栏和标题栏颜色
        primarySwatch: Colors.blue,
        primaryColor: Colors.lightBlue[800],
        accentColor: Colors.cyan[600],
        backgroundColor: Colors.white70
      ),
      home: MyHomePage(title: 'Flutter Demo Home Page'),
    );
  }
}
```

通过以上代码，我们完成了应用主题色调配置，运行即可查看配置后的主题样式及效果，如图 13-1 所示。

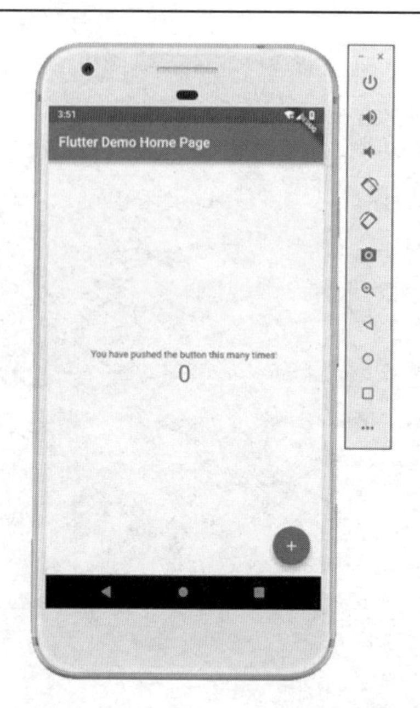

图 13-1　应用主题色调配置效果

建议各位读者多多尝试 ThemeData 组件中的其他属性参数，熟悉它们的作用和配置方法，便于在后续开发中灵活使用。

13.1.2　设置局部主题

有的时候我们不想使用全局主题风格，而是想在某些页面展示不一样的风格，这时该怎么办呢？其实 Flutter 也支持局部主题风格设置。

我们可以使用局部主题覆盖全局主题，此时需要借助 Theme 组件来实现，示例代码如下。

```
// 通过 Theme 组件使局部主题覆盖全局主题
Widget theme2() {
  return Theme(
    data: ThemeData(
      primaryColor: Colors.yellow[800],
      accentColor: Colors.yellow[600],
      backgroundColor: Colors.white,
    ),
    child: Scaffold(
      appBar: AppBar(
        title: Text('Theme'),
```

```
      ),
      body: Text(
        'data',
        style: TextStyle(fontSize: 18, decoration: TextDecoration.none),
      ),
    ),
  );
}
```

通过以上代码可以看出，设置局部主题的方法非常简单：只需在页面最外层包裹一层 Theme 组件，然后在 data 属性里重新配置个性化 ThemeData 组件，最后在 child 属性里放置原来的布局即可。大家可以自己实践，完成部分页面的个性化主题样式设置。

13.1.3 扩展和修改全局主题

有时我们不但要通过固定的局部主题去展现个性化应用风格，还需要动态修改全局主题或扩展全局主题以带来更新鲜的视觉感受。例如，在使用手机 QQ 时，我们可以手动切换全局主题。要想实现这一功能，可以借助 Theme.of(context).copyWith()方法，先来看一个简单的使用示例。

```
// 扩展和修改全局主题
Widget theme3(BuildContext context) {
  return Theme(
    data: Theme.of(context).copyWith(
      // 修改全局主题的 primaryColor 属性
      primaryColor: Colors.white30,
    ),
    child: Scaffold(
      appBar: AppBar(
        title: Text('Theme'),
      ),
      body: Text(
        'data',
        style: TextStyle(
            // 使用主题颜色
            color: Theme.of(context).primaryColor,
            fontSize: 18,
            decoration: TextDecoration.none),
      ),
    ),
```

```
  );
}
```

以上代码的核心是，在想要修改全局主题的地方，将 Theme 组件包裹的 data 属性里的内容用 Theme.of(context).copyWith()方法来扩展，修改部分属性。如果我们想使用全局主题中的一些属性或色调，可以通过 Theme.of(context)来调用，示例代码如下。

```
Container(
  // 调用全局主题的 accentColor 色调
  color: Theme.of(context).accentColor,
  child: Text(
    'A Flutter Theme !',
    style: Theme.of(context).textTheme.title,
  ),
);
```

当然我们可以根据平台的不同使用不同的主题，通过 ThemeData 也可以完成更换皮肤或更换主题的功能。例如，我们想配置两种不同风格的 iOS 系统主题和 Android 系统主题，都可以通过 Flutter 来实现。

13.2 应用国际化

应用国际化就是让应用支持多种语言，国际化处理也称为本地化处理。Flutter 内置的 API 支持国际化处理，当然也可以用官方提供的插件库来实现。本节我们将介绍 Flutter 国际化处理的两种方法，并配合案例进行讲解。

13.2.1 应用国际化简介

要想让应用提供和支持多种语言模式，就要进行国际化处理。在 Flutter 中实现应用国际化一般要配置 MaterialApp 或 WidgetsApp 的国际化属性 localizationsDelegates 和 supportedLocales，并且需要在 pubspec.yaml 中配置一个 flutter_localizations 软件库，该软件库支持 15 种以上不同国家的语言。

下面我们先来看看通过 Flutter 内置 API 实现应用国际化的主要步骤。

首先需要配置 pubspec.yaml，代码如下。

```
dependencies:
  flutter:
```

```
    sdk: flutter
  # 添加国际化包
  flutter_localizations:
    sdk: flutter
```

接下来在需要进行国际化处理的页面导入这个包的相关类，代码如下。

```
import 'package:flutter_localizations/flutter_localizations.dart';
```

然后在 MaterialApp 组件或 WidgetsApp 组件中配置相关的国际化属性 localizationsDelegates 和 supportedLocales，具体代码如下。

```
class LocalizationsSamplesState extends State<LocalizationsSamples> {
  @override
  void initState() {
    super.initState();
  }

  @override
  Widget build(BuildContext context) {
    return MaterialApp(
      localizationsDelegates: [
        GlobalMaterialLocalizations.delegate,
        GlobalWidgetsLocalizations.delegate,
      ],
      supportedLocales: [
        const Locale('en', 'US'), // 英文
        const Locale('zh', 'CN'), // 中文
        // ... 支持其他语言
      ],
      home: getBody(),
    );
  }
...
}
```

supportedLocales 属性用来定义支持的语言列表，Locale 组件用来定义具体语言。Locale 中的参数有两个，分别代表语言和国家。

实现应用国际化还需要用到 Localizations 和 Delegate 两个类，Localizations 类主要用于定义需要翻译的文字对应的各国语言资源，Delegate 类主要负责对各国语言资源进行加载和切换。先来看一下实现 Localizations 类的方法，具体如下。

```
import 'package:flutter/widgets.dart';

class PageLocalizations {
  final Locale locale;
  PageLocalizations(this.locale);

  static Map<String, Map<String, String>> _localizedValues = {
    'en': {
      'task title': 'Flutter Demo',
      'titlebar title': 'Flutter Demo Home Page',
      'click tip': 'You have pushed the button this many times:',
      'inc': 'Increment'
    },
    'zh': {
      'task title': 'Flutter 示例',
      'titlebar title': 'Flutter 示例主页面',
      'click tip': '你一共单击了这么多次按钮：',
      'inc': '增加'
    }
  };

  get taskTitle {
    return _localizedValues[locale.languageCode]['task title'];
  }

  get titleBarTitle {
    return _localizedValues[locale.languageCode]['titlebar title'];
  }

  get clickTop {
    return _localizedValues[locale.languageCode]['click tip'];
  }

  get inc {
    return _localizedValues[locale.languageCode]['inc'];
  }

  static PageLocalizations of(BuildContext context) {
    return Localizations.of(context, PageLocalizations);
```

```
  }
}
```

以上代码定义了需要翻译的文字及对应的语言资源。接下来，我们实现 Delegate 类，代码如下。

```
import 'package:flutter/foundation.dart';
import 'package:flutter/widgets.dart';
import 'package:flutter_samples/samples/pageLocalizations.dart';

class GlobalPagesLocalizations
    extends LocalizationsDelegate<PageLocalizations> {
  const GlobalPagesLocalizations();

  // 是否支持某种语言
  @override
  bool isSupported(Locale locale) {
    return ['en', 'zh'].contains(locale.languageCode);
  }

    // 加载对应的语言资源，自动调用
  @override
  Future<PageLocalizations> load(Locale locale) {
    return new SynchronousFuture<PageLocalizations>(
        new PageLocalizations(locale));
  }

// 重新加载
  @override
  bool shouldReload(LocalizationsDelegate<PageLocalizations> old) {
    return false;
  }

  static GlobalPagesLocalizations delegate = const GlobalPagesLocalizations();
}
```

在以上代码中，LocalizationsDelegate 负责将定义的语言资源进行加载整合，然后在需要切换语言的时候提供资源重载、重新渲染的功能。

实现了以上两个核心类之后，我们就可以调用已完成国际化设置的属性了，具体代码如下。

```
class LocalizationsSamples extends StatefulWidget {
```

```
  @override
  State<StatefulWidget> createState() {
    return LocalizationsSamplesState();
  }
}

class LocalizationsSamplesState extends State<LocalizationsSamples> {
  @override
  void initState() {
    super.initState();
  }

  @override
  Widget build(BuildContext context) {
    return MaterialApp(
      localizationsDelegates: [
        GlobalMaterialLocalizations.delegate,
        GlobalWidgetsLocalizations.delegate,
         // 使用自定义的国际化设置
        GlobalPagesLocalizations.delegate,
      ],
      supportedLocales: [
        const Locale('en', 'US'), // English
        const Locale('zh', 'CN'), // Chinese
        // ... 其他语言支持
      ],
      home: WelcomePage(),
    );
  }
}

class WelcomePage extends StatefulWidget {
  @override
  State<StatefulWidget> createState() {
    return WelcomeState();
  }
}

class WelcomeState extends State<WelcomePage> {
```

```
  @override
  Widget build(BuildContext context) {
    return Scaffold(
      appBar: AppBar(
        title: Text('Localizations'),
        primary: true,
      ),
      body: Column(
        children: <Widget>[
          // 调用已完成国际化设置的属性
          Text(PageLocalizations.of(context).taskTitle,)
        ],
      ),
    );
  }
}
```

运行以上代码，国际化设置前后的效果对比如图 13-2 所示。

图 13-2　国际化设置前后的效果对比

进行国际化设置后，需要在使用对应字段的地方通过 PageLocalizations.of(context).taskTitle 方式进行动态文本资源引用。这样一来，当切换手机语言环境后，应用便会自动显示当前语言环境下的文字。大家可以多增加几个语种进行尝试，以便加深对应用国际化的理解。

13.2.2 使用插件库实现应用国际化

本节介绍通过插件库来实现应用国际化的方法，常用的插件库如 intl 和 flutter_i18n。这两个插件库都是 Flutter 的官方插件库，两者的基本功能是一样的，只不过 flutter_i18n 会自动化执行将 arb 文件转为 dart 文件的操作。

这里以 flutter_i18n 插件库为例。首先我们需要使用 Android Studio 安装 Flutter i18n 插件库，如图 13-3 所示。

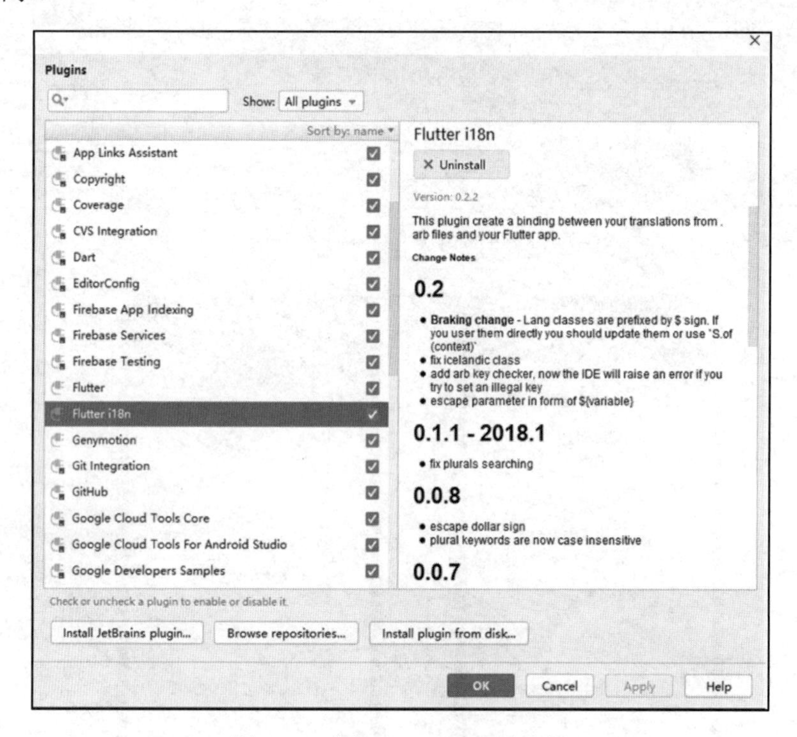

图 13-3　安装 Flutter i18n 插件库

安装完成后会出现一个运行按钮（图 13-4 中被框起来的部分），单击这个按钮将在 lib/generated 目录下自动生成名为 i18n.dart 的文件，如图 13-5 所示。

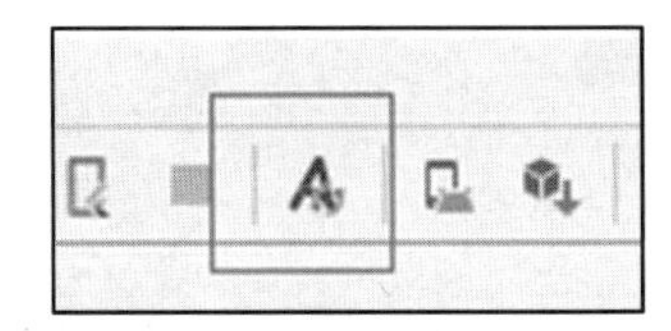

图 13-4　i18n 插件库运行按钮

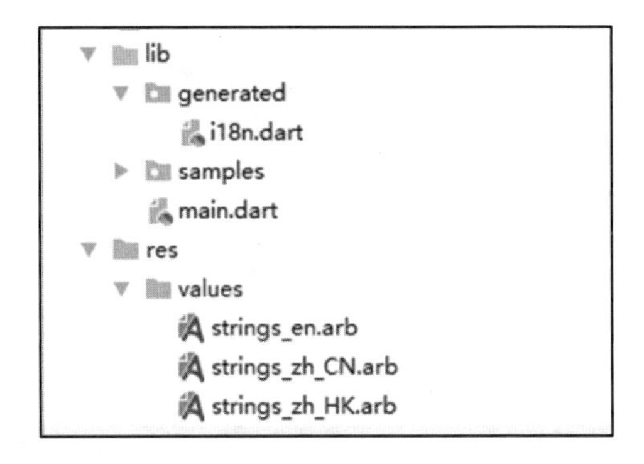

图 13-5　自动生成名为 i18n.dart 的文件

接下来我们来看一下 flutter_i18n 插件库的具体用法。

首先在 res->values 目录处单击鼠标右键，新建需要支持的语言所对应的 arb 文件，选择语言的界面如图 13-6 所示。

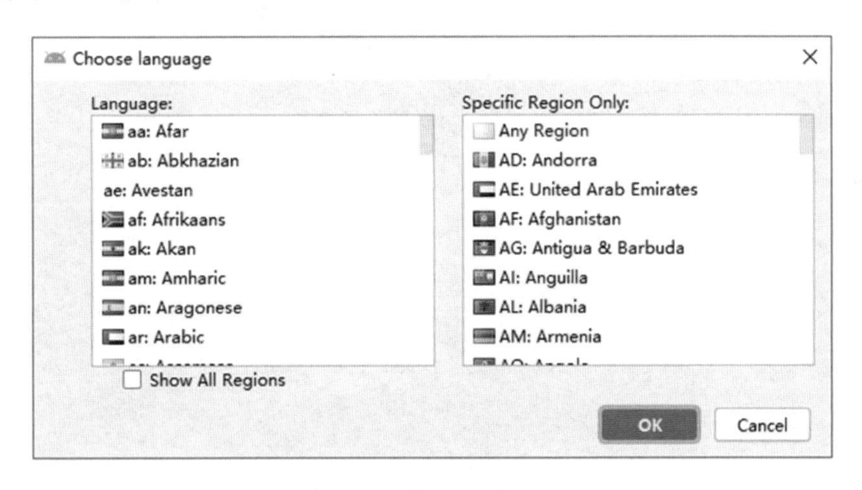

图 13-6　选择语言的界面

本例中我们只新建两种语言的 arb 文件：strings_en.arb 和 strings_zh_CN.arb，分别代表英文和中文。strings_en.arb 和 strings_zh_CN.arb 中的具体内容如下。

```
// strings_en.arb
{
  "appName": "App Name",
  "hello": "Hello $name",
  "title": "My Title"
}
// strings_zn_CN.arb
{
  "appName": "应用名",
  "hello": "你好${name}",
  "title": "我的标题"
}
```

可以看出，这两个文件里配置的都是相关字段的对应翻译，我们也可以通过$符号来传递动态值。

编写完对应语言的 arb 文件后就可以使用插件库自动化生成 Dart 国际化代码了。用鼠标左键单击图 13-4 中的运行按钮，自动生成 i18n.dart 国际化代码文件。这个文件是自动生成的，不可以修改，其中的大致内容如下。

```
// i18n.dart 文件，自动生成
import 'dart:async';

import 'package:flutter/foundation.dart';
import 'package:flutter/material.dart';

class S implements WidgetsLocalizations {
  const S();

  static const GeneratedLocalizationsDelegate delegate =
    GeneratedLocalizationsDelegate();

  static S of(BuildContext context) => Localizations.of<S>(context, S);

  @override
  TextDirection get textDirection => TextDirection.ltr;

  String get appName => "App Name";
  String get title => "My Title";
  String hello(String name) => "Hello $name";
```

```
}

class $zh_HK extends S {
  const $zh_HK();

  @override
  TextDirection get textDirection => TextDirection.ltr;

  @override
  String get appName => "應用名";
  @override
  String get title => "我的標題";
  @override
  String hello(String name) => "妳好${name}";
}

class $en extends S {
  const $en();
}

class $zh_CN extends S {
  const $zh_CN();

  @override
  TextDirection get textDirection => TextDirection.ltr;

  @override
  String get appName => "应用名";
  @override
  String get title => "我的标题";
  @override
  String hello(String name) => "你好${name}";
}

class GeneratedLocalizationsDelegate extends LocalizationsDelegate<S> {
  const GeneratedLocalizationsDelegate();

  List<Locale> get supportedLocales {
    return const <Locale>[
```

```
      Locale("zh", "HK"),
      Locale("en", ""),
      Locale("zh", "CN"),
    ];
  }
… …
```

前面介绍了插件库的安装和其中文件的内容，接下来我们来看一下 flutter_i18n 插件库的使用步骤。首先需要在项目中引用 flutter_i18n 插件库，方法如下。

```
dependencies:
  flutter_i18n: ^0.6.3
```

然后在具体使用的地方导入库，如下。

```
import 'package:flutter_i18n/flutter_i18n.dart';
```

这样一来就可以在项目入口代码里配置并使用国际化类了，具体如下。

```
import 'package:flutter/material.dart';
import 'package:flutter/widgets.dart';
import 'package:flutter_app/generated/i18n.dart';
import 'package:flutter_i18n/flutter_i18n.dart';
import 'package:flutter_localizations/flutter_localizations.dart';

class LocalizationsSamples extends StatefulWidget {
  @override
  State<StatefulWidget> createState() {
    return LocalizationsSamplesState();
  }
}

class LocalizationsSamplesState extends State<LocalizationsSamples> {
      Locale _locale = const Locale('zh', 'CN');

  @override
  void initState() {
    super.initState();
    localeChange = (locale) {
      setState(() {
        _locale = locale;
      });
    };
```

```
  }

  @override
  Widget build(BuildContext context) {
    return MaterialApp(
      localizationsDelegates: [
        GlobalMaterialLocalizations.delegate,
        GlobalWidgetsLocalizations.delegate,
        // 配置 delegate
        S.delegate,
      ],
      supportedLocales: [
        // 支持的语言
        const Locale('en', ''), // 英文
        const Locale('zh', 'CN'), // 中文
        const Locale("zh", "HK"),
        // ... 支持其他语言
      ],
      // 也可以指定一种默认语言
      localeResolutionCallback:
          S.delegate.resolution(fallback: const Locale('en', '')),
      home: WelcomePage(),
    );
  }
}

class WelcomePage extends StatefulWidget {
  @override
  State<StatefulWidget> createState() {
    return WelcomeState();
  }
}

class WelcomeState extends State<WelcomePage> {
  @override
  Widget build(BuildContext context) {
    return Scaffold(
      appBar: AppBar(
        title: Text('Localizations'),
```

```
        primary: true,
      ),
      body: Column(
        children: <Widget>[
          // 调用已完成国际化设置的属性
          Text(
            S.of(context).title,
          )
        ],
      ),
    );
  }
}
```

可以看到，使用插件库实现应用国际化的方法和前面介绍的通过内置 API 实现应用国际化的方法是基本一样的，只不过很多重复的代码逻辑可以自动化生成，更加方便。如果我们想在某个地方动态切换语言，例如按下某个按钮将应用语言由中文转为英文，可以如下操作。

```
FlutterI18n.refresh(context, Locale('en', ''));
```

除了以上这个修改全局语言的方法，我们还可以指定修改某个字段的语言，方法如下。

```
FlutterI18n.translate(buildContext, "your.key")
FlutterI18n.plural(buildContext, "select", 0)
```

以上我们介绍了两种应用国际化的实现方法，这里推荐使用第二种方法，因为通过插件库实现应用国际化比较方便、快捷。当然，如果应用并不涉及多语言使用场景，也可以不进行国际化处理。

第 14 章 Flutter 数据共享与传递

在使用 Flutter 进行开发的过程中，可能需要进行页面之间数据的共享、传递，以及对全局组件、状态进行管理和监听。在本章中，我们将详细介绍 Flutter 中数据共享与传递的方法：通过 InheritedWidget 组件、通过 ScopedModel 库、通过 Redux 库、通过 EventBus 库。同时也会介绍一些与数据交互相关的插件库。

14.1 InheritedWidget 组件

在 Android 开发中，可以通过静态常量、变量、全局对象来共享数据，也可以通过数据库进行数据存储和共享。Flutter 中则是以另一种方式来实现全局数据共享与传递的，例如，我们可以使用 InheritedWidget 组件。

通过 InheritedWidget 组件实现数据共享与传递的方式有点类似于 Flutter 中获取全局主题某个属性的方式。和单例模式的用法很像，获取全局主题某个属性的方式如下。

```
// 获取全局主题某个属性
Theme.of(context).primaryColor;
... ...
@override
Widget build(BuildContext context) {
  return Text(
'Example',
    // 获取某个属性
    style: Theme.of(context).textTheme.title,
  );
}
```

通过以上代码可以看到，调用 Theme.of(context)来获取一个全局对象即可通过这个全局对象获取它的属性。

还有一个典型的使用 InheritedWidget 组件来实现数据共享与传递的例子，即使用 MediaQuery 组件。MediaQuery 组件继承自 InheritedWidget 组件，可借助 InheritedWidget 的特点来实现数据共享与传递。MediaQuery 的调用方式也类似于 Theme 全局对象，例如通过 MediaQuery 来获取屏幕宽度的语句是 MediaQuery.of(context).size.width，即先获取全局对象，再获取该对象的属性。

接下来我们来看一个官方给出的通过 InheritedWidget 组件来实现的数据共享与传递的示例，代码如下。

```
class FrogColor extends InheritedWidget {
  const FrogColor({
    Key key,
    @required this.color,
    @required Widget child,
  }) : assert(color != null),
       assert(child != null),
       super(key: key, child: child);

  final Color color;

  static FrogColor of(BuildContext context) {
    return context.inheritFromWidgetOfExactType(FrogColor) as FrogColor;
  }

  @override
  bool updateShouldNotify(FrogColor old) => color != old.color;
}

// 下面是使用方式
... ...
@override
  Widget build(BuildContext context) {
    return Scaffold(
        appBar: AppBar(title: Text('File Samples'), primary: true),
        // 设置颜色
        body: FrogColor(
```

```
            color: Colors.teal,
            child: Text("文件操作"),
          )); // Text("文件操作")
    }

// 获取颜色值的操作类似于 MediaQuery 用法
FrogColor.of(context).color.value;
```

通过上面的代码可以看到，我们需要先通过 FrogColor.of(context)来获取全局对象，实际上调用的方法是 context.inheritFromWidgetOfExactType(Type)，通过这个方法可以直接返回对象，也可以返回值。

其中 updateShouldNotify 方法的主要作用是判断是否需要更新数据，当新的值和原有的值不一样时就需要更新对象里的数据。

通过 InheritedWidget 组件只能获取数据，不能修改数据，这是它的一个缺点。想要修改数据，我们需要进行功能扩充。

14.2　ScopedModel 库

由于通过 InheritedWidget 组件来实现数据共享与传递有其自身的限制，所以我们可以采用另外一种方式来更方便地实现全局数据共享、传递、监听，即借助 ScopedModel 库。

ScopedModel 库的作用是进行状态管理、数据管理。假设我们进入某个详情页面，并进行了评论、点赞、更改某些状态的操作，之后再返回前一个页面，这时页面中的数据也应该得到更新。在这种情况下，我们可以使用 ScopedModel 库来实现数据的共享和传递。

ScopedModel 库使用观察者模式实现数据状态管理，内部使用 InheritedWidget 组件进行数据共享、传递。接下来我们通过一个示例来查看 ScopedModel 库的具体使用方法，以下代码实现的功能是，在第二个页面通过特定手势增加数值，返回第一个页面后同步更新数值。

首先我们需要在 pubspec.yaml 里配置并引用 ScopedModel 库，代码如下。

```
dependencies:
  scoped_model: ^1.0.1
```

然后在使用的地方导入 scoped_model.dart 类。

```
import 'package:scoped_model/scoped_model.dart';
```

导入完毕，我们需要定义一个 Model，通过这个 Model 来存储数据，代码如下。

```
import 'package:flutter/widgets.dart';
import 'package:scoped_model/scoped_model.dart';

// 定义Model，用来存储数据
class CounterModel extends Model {
  // 定义属性
  int _counter = 0;
  // 获取值的方法
  int get counter => _counter;
  // 定义改变值的方法
  void increment() {
    _counter++;
    // 刷新数据
    notifyListeners();
  }
  // 也可以通过以下方法获取对象，调用里面的属性和方法
  static CounterModel of(BuildContext context) {
    return ScopedModel.of<CounterModel>(context);
  }
}
// 想实现数据监听父页面，最外层需用 ScopedModel 包裹，然后自定义 Model
// 在应用入口处用 ScopedModel 包裹
void main() {
  return runApp(
      ScopedModel<CounterModel>(model: CounterModel(), child: ShowApp()));
}
```

自定义 Model 后，接下来就可以在需要使用共享数据的页面调用 Model 了。先来看一下本例中第一个页面如何编写，即如何在第一个页面获取数据，代码如下。

```
// 子页面获取数据
import 'package:flutter/material.dart';
import 'package:flutter/widgets.dart';
import 'package:scoped_model/scoped_model.dart';

import 'countermodel.dart';
import 'scoped_detail_samples.dart';

class ScopedSamples extends StatefulWidget {
```

```
  @override
  State<StatefulWidget> createState() {
    return ScopedSamplesState();
  }
}

class ScopedSamplesState extends State<ScopedSamples> {
  @override
  void initState() {
    super.initState();
  }

  @override
  Widget build(BuildContext context) {
    // 实例化获取数据
    final _model = CounterModel.of(context).counter;
    return Scaffold(
      appBar: AppBar(title: Text('Scoped Samples'), primary: true),
      body: Column(
        children: <Widget>[
          // 方式一
          ScopedModelDescendant<CounterModel>(builder: (context, child, model) {
            return Text('${model.counter}');
          }),
          // 方式二
          Text('$_model')
        ],
      ),
      floatingActionButton: FloatingActionButton(
        onPressed: () {
          Navigator.push(context, MaterialPageRoute(builder: (context) {
            return ScopedDetailSamples();
          }));
        },
      ),
    );
  }
}
```

我们可以看到，在第一个页面中获取数据的方式有两种，非常方便，不过第二种看起来更简单一些。

接下来我们编写目标跳转页面的代码，即第二个页面的代码，实现在跳转后的页面获取数据的功能，具体如下。

```
import 'package:flutter/material.dart';
import 'package:flutter/widgets.dart';
import 'package:scoped_model/scoped_model.dart';

import 'countermodel.dart';

class ScopedDetailSamples extends StatefulWidget {
  @override
  State<StatefulWidget> createState() {
    return ScopedDetailSamplesState();
  }
}

class ScopedDetailSamplesState extends State<ScopedDetailSamples> {
  @override
  void initState() {
    super.initState();
  }

  @override
  Widget build(BuildContext context) {
    // 实例化获取数据
    final _model = CounterModel.of(context).counter;
    return Scaffold(
      appBar: AppBar(title: Text('Scoped Samples'), primary: true),
      // 获取数据
      body: Text('$_model'),
      floatingActionButton: FloatingActionButton(
        // 调用对象的 increment 方法
        onPressed: CounterModel.of(context).increment,
        tooltip: 'Increment',
        child: Icon(Icons.add),
      ),
    );
```

```
  }
}
```

通过以上代码，我们就实现了使用 ScopedModel 库使两个页面的数据同步更新的效果。可以看出，相对于使用 InheritedWidget 组件，借助 ScopedModel 库不仅可以方便地进行数据共享，还可以对数据进行动态修改，灵活度更高。

14.3　Redux 库

Redux 库和 ScopedModel 库的作者是同一个人，不过相对于 ScopedModel 库，Redux 库使用起来麻烦一些。Redux 库主要由 State、Action、Reduxer、Store、StoreConnector 几部分组成。

- State：相当于 ScopedModel 库里的 Model，用于定义对象和属性。
- Action：用于定义操作 State 的方法。
- Reduxer：将 Action 和 State 进行关联匹配的中间件。
- Store：初始化实体对象。
- StoreConnector：用于绑定、更新数据。

关于 Redux 库的用法，我们依然使用官方提供的增加数字的例子来进行讲解和演示。首先需要在 pubspec.yaml 里配置并引用 Redux 库，代码如下。

```
dependencies:
  flutter_redux: ^0.5.3
```

接着在使用的地方导入 flutter_redux.dart 类。

```
import 'package:flutter_redux/flutter_redux.dart';
```

然后定义一个 State 实体对象，代码如下。

```
// 定义 State 实体对象，内含属性和构造方法
class AppState {
  int counter;

  AppState(this.counter);
}
```

这个 AppState 类可以单独写在一个文件里，也可以写在内部类里。接下来定义操作 State 实体类的方法集合 Actions（Action 方法的集合形式），代码如下。

```
// 定义Actions，多个
enum Actions { Increment, Decrement }
```

然后定义 Reduxer 中间件，用于关联 Actions 和 AppState，匹配两者的数据，代码如下。

```
// 定义Reduxer中间件reducer，关联Actions和AppState
AppState reducer(AppState state, action) {
  if (action == Actions.Increment) {
    return new AppState(state.counter + 1);
  } else if (action == Actions.Decrement) {
    return new AppState(state.counter - 1);
  }
  return state;
}
```

有了这几个类之后，就可以开始构建 Store 了。

```
// 构建Store，初始化实体对象并赋值
final store = Store(reducer, initialState: AppState(0));
```

为了方便，我们把上述几个核心部分写在一个单独的 dart 文件里，具体如下。

```
// dart文件名：redux_app.dart
// 定义State实体对象，内含属性和构造方法
class AppState {
  int counter;
  AppState(this.counter);
}
// 定义Actions，可以定义多个
enum Actions { Increment, Decrement }
// 定义Reduxer中间件reducer，关联Actions和AppState
AppState reducer(AppState state, action) {
  if (action == Actions.Increment) {
    return new AppState(state.counter + 1);
  } else if (action == Actions.Decrement) {
    return new AppState(state.counter - 1);
  }
  return state;
}
// 构建Store，初始化实体对象并赋值
final store = Store(reducer, initialState: AppState(0));
```

接下来需要在入口类 main.dart 里包裹一个 StoreProvider，将 Store 绑定上去进行全局监听、

数据共享，代码如下。

```
void main() => runApp(ShowApp());

class ShowApp extends StatelessWidget {
  @override
  Widget build(BuildContext context) {
    return StoreProvider<AppState>(
      store: store,
      child: MaterialApp(
        title: 'Flutter Demo',
        theme: ThemeData(
          primarySwatch: Colors.teal,
        ),
        home: ShowAppPage(),
        routes: <String, WidgetBuilder>{
          '/buttonpage': (BuildContext context) => ButtonSamples(),
          '/routepage': (BuildContext context) => RouteSamples(),
        },
      ),
    );
  }
}
```

以上方法和 ScopedModel 库用法很像，都需要将入口处的组件进行包裹。使用 Redux 库时，我们要在相应页面使用 StoreConnector 包裹组件，以实现数据共享。StoreConnector 的作用和 ScopedModel 里的 ScopedModelDescendant 基本一致，大家可以对比学习。接下来我们来看一下如何在子页面监听、共享数据，代码如下。

```
// 绑定数据有两种方式
// 第一种：通过 StoreProvider.of<AppState>(context).state 来获取属性值
int counter = StoreProvider.of<AppState>(context).state.counter;
... ...
body: Text(
        '$counter',
        style: TextStyle(fontSize: 20),
      ),
// 第二种：用 StoreConnector 包裹组件
@override
  Widget build(BuildContext context) {
```

```
    return Scaffold(
      appBar: AppBar(title: Text('Redux Samples'), primary: true),
      body: StoreConnector<AppState, String>(
        // 定义数据转换方式
        converter: (store) => store.state.counter.toString(),
        // 定义组件布局，绑定数据，第二个参数对应 converter 返回的数据
        builder: (context, counter) {
          return Text(
            counter,
            style: TextStyle(fontSize: 20),
          );
        },
      ),
      floatingActionButton: StoreConnector<AppState, VoidCallback>(
        converter: (store) {
          return () {
            // store.dispatch 用于执行 Actions 里的方法
            return store.dispatch(reduxApp.Actions.Increment);
          };


        },
        // 简化为 lambda 表达式：
        // converter: (store) => () => store.dispatch(Actions.Decrement),
        builder: (context, callback) {
          return FloatingActionButton(
            // 单击 FloatingActionButton 执行 converter 返回的操作
            onPressed: callback,
            child: Icon(Icons.add),
          );
        },
      ),
    );
  }
}
```

通过以上代码，我们就完整实现了用 Redux 库进行数据监听与共享的功能。该方法不难理解，但相较 ScopedModel 库使用方法更麻烦一些。大家可以对比学习，根据自己的实际需求合理选择。

14.4 EventBus 库

最后我们看一下通过 EventBus 库来实现数据监听、共享与传递的方法。

对于 Android 平台开发者来说，EventBus 库并不陌生，EventBus 在 Android 平台上是一个集事件总线、通信、数据共享、监听功能于一身的第三方库。后来，有人实现了 Flutter 平台的 EventBus 库，用于实现数据监听、共享、传递，大家可以在 Dart Pub 里搜索。

EventBus 库是一个解耦应用程序的简单事件总线，基于发布/订阅模式，有发布者、订阅者两个角色。以 MVC 模式为例，MVC 模式中有模型（Model）、视图（View）、控制器（Controller）三个角色，主要用于将业务逻辑、数据、界面分离解耦。当只有一组 MVC 关系时，很容易进行数据通信，如图 14-1 所示。

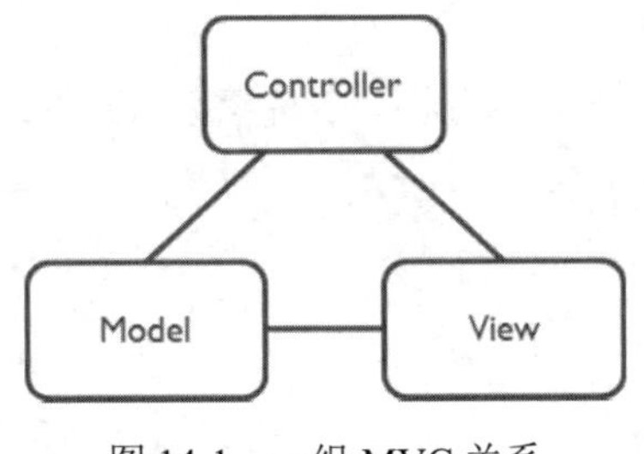

图 14-1 一组 MVC 关系

但是当有多组 MVC 关系时，如图 14-2 所示，多组控制器之间进行通信会非常麻烦和复杂，因为耦合度高，管理起来也非常不方便。

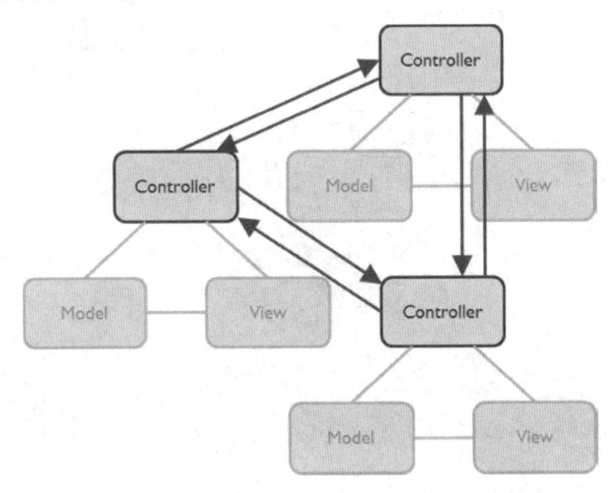

图 14-2 多组 MVC 关系

如果能通过 EventBus 库事件总线管理多组控制器，耦合度将大大降低，如图 14-3 所示。

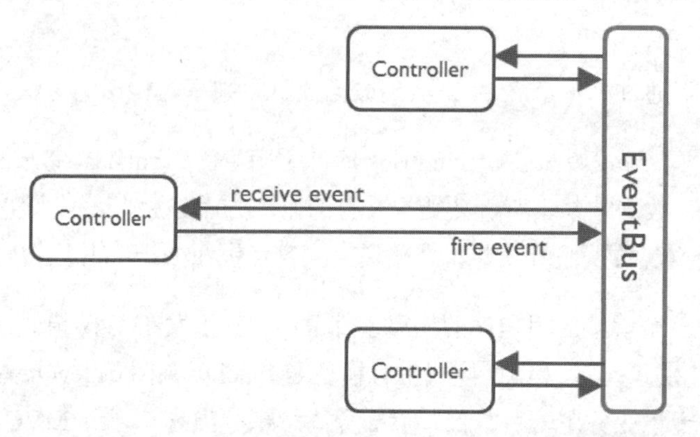

图 14-3　通过 EventBus 库事件总线管理多组控制器

在了解了 EventBus 库的特点之后，我们来看一下 EventBus 库的具体用法。EventBus 库相对于前面介绍的两个库更加容易理解和使用。首先需要添加 EventBus 依赖库。

```
dependencies:
  event_bus: ^1.1.0
```

然后在使用的地方导入 event_bus.dart 类。

```
import 'package:event_bus/event_bus.dart';
```

接下来需要创建一个事件总线，代码如下。

```
import 'package:event_bus/event_bus.dart';
// 实例化 EventBus 库，创建一个事件总线
EventBus eventBus = EventBus();
```

然后再来定义 Event 事件，也可以称之为实体类，代码如下。

```
// 定义 Event 事件，即要传递和共享的实体类
class UserEvent {
  String name;

  UserEvent(this.name);
}

// 可以定义多个实体类
class BookEvent {
  String bookName;
```

```
  BookEvent(this.bookName);
}
```

有了 EventBus 库的 Event 事件后，需要注册这个事件的观察者，代码如下。

```
eventBus.on<UserEvent>().listen((event) {
  // 所有类型为 UserEvent 或其子类的事件、数据都可以被监听
  print(event.name);
});

eventBus.on().listen((event) {
  // 监听所有事件
  print(event.runtimeType);
});
```

上面代码的功能是通过监听获取数据的更新，当有数据发送过来时，观察者就会立即收到，进行页面更新或其他相关操作。

有了观察者，我们来看一下事件如何发布，具体代码如下。

```
User myUser = User('Mickey');
eventBus.fire(UserEvent(myUser));
```

如果想要取消事件订阅，即不再继续进行数据监听，可以使用下面的代码实现。

```
StreamSubscription userSubscription = eventBus.on<UserEvent>().listen((event) {
  print(event.name);
});

userSubscription.cancel();
```

以上介绍了 EventBus 库使用过程中比较重要的几个步骤，接下来我们通过一个示例来看一下 EventBus 库的具体用法。这里我们依然实现两个页面，第一个页面用于接收、监听数据，第二个页面用来发送数据。

首先新建一个类，这里命名为 event_bus.dart，这个类用于定义 Event 事件，即实体类，代码如下。

```
import 'package:event_bus/event_bus.dart';

// 实例化 EventBus 库，创建一个事件总线
EventBus eventBus = EventBus();

// 定义 Event 事件，即要传递和共享的实体类
```

```
class UserEvent {
  String name;

  UserEvent(this.name);
}

// 可以定义多个实体类
class BookEvent {
  String bookName;

  BookEvent(this.bookName);
}
```

定义好实体类后，我们来编写第一个页面的代码。在第一个页面上，要完成注册、监听事件和数据，具体代码如下。

```
class EventBusSamplesState extends State<EventBusSamples> {
  var name = '初始数据';

  @override
  void initState() {
    super.initState();
    // 注册和监听发送来的 UserEven 类型事件、数据
    eventBus.on<UserEvent>().listen((UserEvent event) {
      setState(() {
        name = event.name;
      });
    });
  }

  @override
  Widget build(BuildContext context) {
    return Scaffold(
      appBar: AppBar(title: Text('EventBus Samples'), primary: true),
      // 绑定数据
      body: Text(
        '$name',
        style: TextStyle(fontSize: 20),
      ),
      floatingActionButton: FloatingActionButton(
```

```
        onPressed: () {
          Navigator.push(context, MaterialPageRoute(builder: (context) {
            return EventBusDetailSamples();
          }));
        },
        tooltip: '跳转',
        child: Icon(Icons.add),
      ),
    );
  }
}
```

通过以上代码可以看到，在页面初始化时就注册了事件观察者，用于监听数据变化。接下来我们来编写第二个页面的代码，这个页面用来发送事件、数据，其他页面的观察者可以接收到，具体代码如下。

```
class EventBusDetailSamplesState extends State<EventBusDetailSamples> {
  var name = '初始数据'
  @override
  void initState() {
    super.initState();
    // 注册和监听发送来的 UserEven 类型事件、数据
    eventBus.on<UserEvent>().listen((UserEvent event) {
      setState(() {
        name = event.name;
      });
    });
  }

  @override
  Widget build(BuildContext context) {
    return Scaffold(
      appBar: AppBar(title: Text('EventBus Samples'), primary: true),
      body: Text(
        '$name',
        style: TextStyle(fontSize: 20),
      ),
      floatingActionButton: FloatingActionButton(
        onPressed: () {
          // 发送事件、数据
```

```
            eventBus.fire(UserEvent('Tom'));
          },
          child: Icon(Icons.add),
        ),
      );
    }
}
```

通过以上代码可以看到，EventBus 库适用于全局数据通信场景，不适合持久化保存数据，因此可以搭配 ScopedModel 库使用。如果想持久化保存数据，需要使用数据库一类的持久化方式对数据进行本地化或网络化存储。

第 15 章

Flutter 与原生 API 交互及插件库开发

在使用 Flutter 进行开发的过程中，可能有各种各样的需求、技术方案无法通过现有的组件来实现，这个时候我们需要开发插件库（实际上就是调用原生平台的 API）来实现相应的交互功能。本章我们将学习 Flutter 与原生 API 交互的方法，插件库开发方式，以及常见插件库的用法。

15.1 Flutter 与原生 API 交互

在开发应用时，可能会涉及 Flutter 与原生平台的 API 进行交互调用的场景，如获取手机电池电量、打开摄像头等。那么这种交互如何实现呢，本节将具体讲解。

15.1.1 交互简介

在开发过程中，如果遇到某些功能无法通过 Flutter 实现，可以选择使用第三方插件库。如果没有可用的第三方插件库，就需要自己编写与原生 API 交互的逻辑。当然，Flutter 也支持混合开发，即在原生 App 里加入 Flutter 页面或在 Flutter 应用里加入原生页面。

Flutter 与原生 API 交互的核心是通过 MethodChannel 组件进行传值和方法调用，其原理如图 15-1 所示。

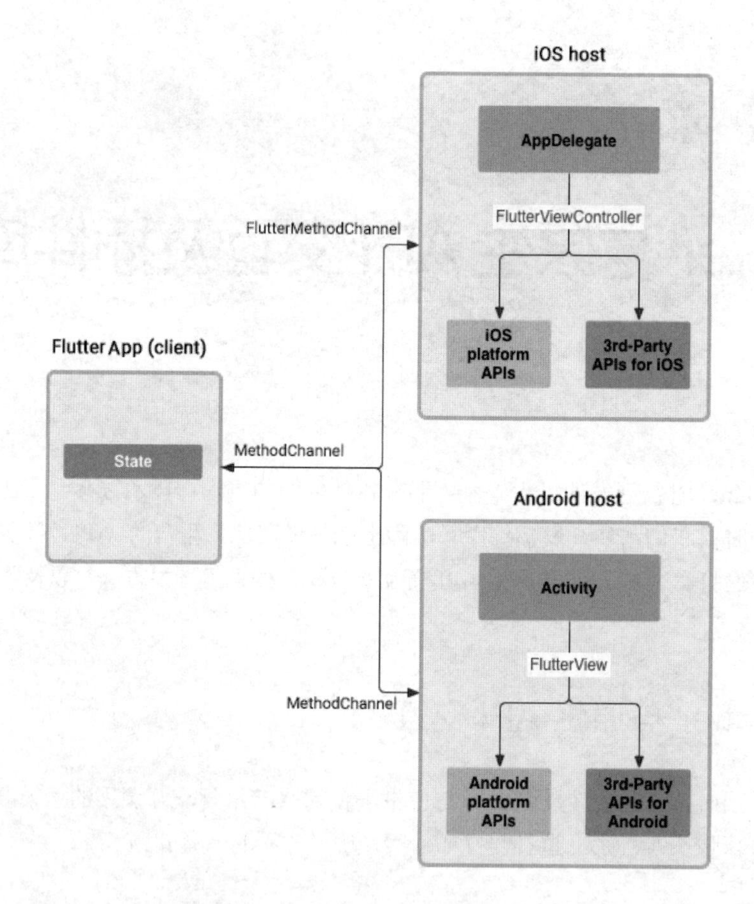

图 15-1　Flutter 与原生 API 交互原理

从图 15-1 中可以看出，Flutter 与 Android 和 iOS 主要是通过 MethodChannel 组件进行交互、传值、调用的，并且支持双向调用与通信交互。Flutter 插件库其实就是通过 Flutter 与原生 API 进行交互而实现的，例如我们可以通过一个 Flutter 相机插件库来实现调用原生相机拍照的功能，这些都是 Flutter 与原生 API 交互的应用场景。

我们可以把 Flutter 端看成客户端，把对应的原生 Android 端和 iOS 端看成服务器端，双方的交互是通过消息发送来实现的。Android 端和 iOS 端通过 MethodChannel 发送消息给 Flutter 端；而 Flutter 端分别通过 MethodChannel 和 FlutterMethodChannel 发送数据、消息给 Android 端和 iOS 端，这样就实现了双向通信。

下面我们以 Android 平台为例，介绍 Flutter 与原生 API 交互的几种实现方法。

15.1.2 调用原生 API

在编写 Flutter 调用原生 API 的程序时，建议大家将 Android Studio 或 IntelliJ Idea 作为开发工具，不建议使用 VSCode。因为编写程序时会涉及 Android 原生代码，VSCode 是不支持的，但 Android Studio 能提供比较完善的支持。

那么如何在 Flutter 中调用原生 API 并传递参数，然后从 Android 原生平台返回值呢？我们来看一个 Flutter 官方给出的简单例子：调用原生 API 获取电池电量信息并返回。首先使用 Android Studio 创建 Flutter 项目，结构如图 15-2 所示。

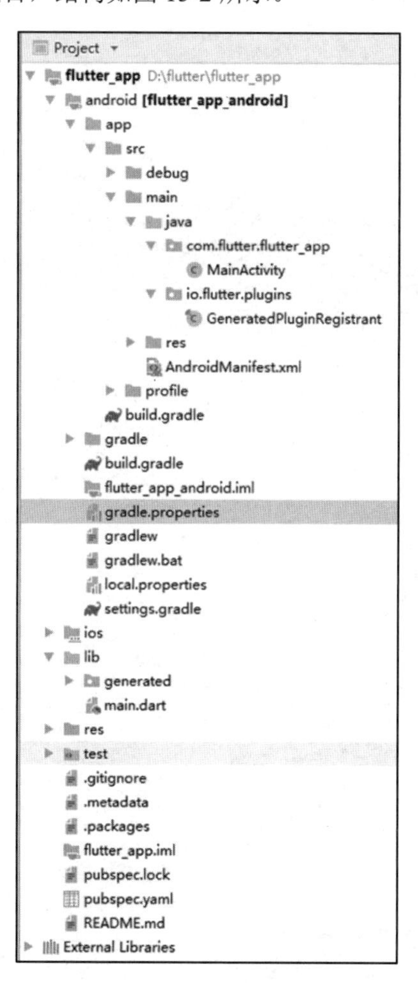

图 15-2 Flutter 项目结构

项目的主要逻辑代码都是写在 MainActivity 里的，具体如下。需要注意的是，我们要定义通信所需的唯一标识符 CHANNEL，原生 Android 端定义的 CHANNEL 名称要知 Flutter 端定义的 CHANNEL 名称一致，这样才可以匹配通信。

```
package com.flutter.flutter_app;

import android.content.ContextWrapper;
import android.content.Intent;
import android.content.IntentFilter;
import android.os.BatteryManager;
import android.os.Build;
import android.os.Bundle;

import io.flutter.app.FlutterActivity;
import io.flutter.plugin.common.MethodCall;
import io.flutter.plugin.common.MethodChannel;
import io.flutter.plugins.GeneratedPluginRegistrant;

public class MainActivity extends FlutterActivity {
    // 定义 CHANNEL 名称
    private static final String CHANNEL = "samples.flutter.io/battery";

    @Override
    protected void onCreate(Bundle savedInstanceState) {
        super.onCreate(savedInstanceState);
        // Flutter 插件库注册
        GeneratedPluginRegistrant.registerWith(this);
        // 使用 MethodChannel 进行通信
        new MethodChannel(getFlutterView(), CHANNEL).setMethodCallHandler(
                new MethodChannel.MethodCallHandler() {
             @Override
             public void onMethodCall(MethodCall call, MethodChannel.Result result) {
                // 使用 MethodCall 进行方法名匹配
                if (call.method.equals("getBatteryLevel")) {
                  // 获取电池电量信息
                  int batteryLevel = getBatteryLevel();
                  if (batteryLevel != -1) {
                    // 通过 MethodChannel.Result 返回结果给 Flutter 客户端
                    result.success(batteryLevel);
```

```
                } else {
                 result.error("UNAVAILABLE", "Battery level not available.", null);
                }
                 } else {
                 result.notImplemented();
                  }
                }
              });
    }
    // 编写原生Android端获取电池电量信息的方法
    private int getBatteryLevel() {
        int batteryLevel = -1;
        if (Build.VERSION.SDK_INT >= Build.VERSION_CODES.LOLLIPOP) {
            BatteryManager batteryManager = (BatteryManager) getSystemService(BATTERY
_SERVICE);
            batteryLevel = batteryManager.getIntProperty(BatteryManager.BATTERY_PROPE
RTY_CAPACITY);
        } else {
            Intent intent = new ContextWrapper(getApplicationContext()).
                    registerReceiver(null, new IntentFilter(Intent.ACTION_BATTERY_CHA
NGED));
            batteryLevel = (intent.getIntExtra(BatteryManager.EXTRA_LEVEL, -1) * 10
0) /
                    intent.getIntExtra(BatteryManager.EXTRA_SCALE, -1);
        }
        return batteryLevel;
    }
}
```

上述代码的核心是 MethodChannel 里的 onMethodCall 方法，核心的交互通信功能就是在这个方法里完成的。我们着重看一下 onMethodCall 方法里的两个参数 MethodCall 和 MethodChannel.Result，MethodCall 的结构代码如下。

```
public final class MethodCall {
    public final String method;
    public final Object arguments;
    ... ...
}
```

可以看到，MethodCall 参数有两个属性：一个是 method，用于将调用的方法名传递到 Flutter 端（客户端）；另一个是 arguments，用于传递参数给原生 Android 端（服务器端），这个参数就是 Flutter 传递过来的要调用的方法附带的参数。例如我们想调用原生平台打开相册图片的方法，method 设置的是打开相册图片的方法名，arguments 设置的是要打开的图片的路径。

接下来我们来看另外一个参数 MethodChannel.Result。这个参数主要用于将原生 Android 端获取的数据传递给 Flutter 端，其结构代码如下。

```
public interface Result {
    void success(@Nullable Object var1);

    void error(String var1, @Nullable String var2, @Nullable Object var3);

    void notImplemented();
}
```

可以看到，MethodChannel.Result 结构中有三个回调方法：success，表示成功回调，可以通过这个方法回调正确数据或结果给客户端；error，错误回调，可以通过这个方法回调错误信息给客户端；notImplemented，未实现的方法。

原生 Android 端的代码逻辑写好了，接下来我们来编写 Flutter 端的代码逻辑。Flutter 端主要负责传递要调用的方法名和参数值，以及获取原生 Android 端返回的结果，具体代码如下。

```
import 'package:flutter/material.dart';
import 'package:flutter/services.dart';
import 'package:flutter/widgets.dart';

class NativeSamples extends StatefulWidget {
  @override
  State<StatefulWidget> createState() {
    return NativeSamplesState();
  }
}

class NativeSamplesState extends State<NativeSamples> {
  // 创建 MethodChannel
  static const platform = const MethodChannel('samples.flutter.io/battery');
  // 定义电池电量信息变量
  String _batteryLevel = 'Unknown battery level.';

  @override
  void initState() {
```

```
    super.initState();
  }

  @override
  Widget build(BuildContext context) {
    return Scaffold(
      appBar: AppBar(
        title: Text('Flutter with Native'),
        primary: true,
      ),
      body: Column(
        children: <Widget>[
          // 获取电池电量信息
          RaisedButton(
            child: Text('Get Battery Level'),
            onPressed: _getBatteryLevel,
          ),
          // 显示电池电量信息
          Text(_batteryLevel),
        ],
      ),
    );
  }
  // 调用在原生 Android 端获取电池电量信息的方法
  Future<Null> _getBatteryLevel() async {
    String batteryLevel;
    try {
      // 通过 invokeMethod 进行反射调用，获取原生方法名，获取返回值
      final int result = await platform.invokeMethod('getBatteryLevel');
      batteryLevel = 'Battery level at $result % .';
    } on PlatformException catch (e) {
      batteryLevel = "Failed to get battery level: '${e.message}'.";
    }
    // 更新返回值
    setState(() {
      _batteryLevel = batteryLevel;
    });
  }
}
```

以上为 Flutter 端逻辑代码，有了原生 Android 端和 Flutter 端的交互逻辑代码， Flutter 与原生 API 便能实现交互。

刚刚我们实现了 Flutter 官方给出的简单示例，接下来再来实现一个稍微复杂的例子：在 Flutter 中按下按钮，传递参数，调用原生 Android 端的权限申请方法并返回状态。步骤与前面的例子基本相同，因此不再赘述，我们直接来看 Flutter 端的代码逻辑，具体如下。

```
// 判断是否有权限，无权限就主动申请权限
Future<Null> _requestPermission() async {
  bool hasPermission;
  try {
    // 传递参数，键值对形式
    hasPermission =
        await platform.invokeMethod('requestPermission', <String, dynamic>{
      'permissionName': 'WRITE_EXTERNAL_STORAGE',
      'permissionId': 0,
    });
  } on PlatformException catch (e) {
    hasPermission = false;
  }

  setState(() {
    _hasPermission = hasPermission;
  });
}
```

通过以上代码可以看到，在调用原生方法时，除了要声明方法名，还要传递一些必要的参数给原生 API，传递时可以传递多个。传递参数主要通过 invokeMethod 方法实现，invokeMethod 的构造方法如下。

```
Future<T> invokeMethod<T>(String method, [ dynamic arguments ]) async{
  ... ...
}

// 使用示例
_channel.invokeMethod('play', <String, dynamic>{
        'song': song.id,
        'volume': volume,
}
```

可以看到，invokeMethod 方法中的第一个参数是要调用的原生方法名，另一个参数是动态传递给原生方法的相关参数（如果没有参数也可以不传）。再来看一下原生 Android 端的代码逻辑，具体如下。

```
public class MainActivity extends FlutterActivity {
    private static final String CHANNEL = "samples.flutter.io/battery";

    @Override
    protected void onCreate(Bundle savedInstanceState) {
        super.onCreate(savedInstanceState);
        // Flutter 插件库注册
        GeneratedPluginRegistrant.registerWith(this);

        new MethodChannel(getFlutterView(), CHANNEL).setMethodCallHandler(
                new MethodChannel.MethodCallHandler() {
            @Override
            public void onMethodCall(MethodCall call, MethodChannel.Result result) {
                switch (call.method) {
                    case "requestPermission":
                        // 申请权限，获取参数
                        final String permissionName = call.argument("permissionName");
                        final int permissionId = call.argument("permissionId");
                        // 调用申请权限的方法
                        boolean hasPermission = requestPermission();
                        System.out.println(permissionName + "    " + permissionId);
                        // 回调返回结果，传递给 Flutter 端
                        result.success(hasPermission);
                        break;
                    default:
                        result.notImplemented();
                    }
                }
            });
    }

    // 请求权限
    private boolean requestPermission() {
        if (Build.VERSION.SDK_INT >= Build.VERSION_CODES.M) {
```

```
            if (checkSelfPermission(Manifest.permission.WRITE_EXTERNAL_STORAGE) == Pa
ckageManager.PERMISSION_DENIED) {
                requestPermissions(
                        new String[]{Manifest.permission.WRITE_EXTERNAL_STORAGE},
                        0);
                return false;
            } else {
                return true;
            }
        }
        return true;
    }

    @Override
    public void onRequestPermissionsResult(int requestCode, String[] permissions, int
[] grantResults) {
        super.onRequestPermissionsResult(requestCode, permissions, grantResults);
        if (requestCode == 0) {
            if (grantResults[0] == PackageManager.PERMISSION_GRANTED) {
                Toast.makeText(this, "权限已申请", Toast.LENGTH_SHORT).show();
            } else {
                Toast.makeText(this, "权限已拒绝", Toast.LENGTH_SHORT).show();
            }
        }
    }
}
```

通过以上代码，我们就实现了一个复杂的 Flutter 与原生 API 交互的示例，并且传递了一些参数。大家可以尝试按照这个思路去编写 Flutter 与原生 API 交互的功能代码。

15.1.3　原生 API 调用 Flutter API

从本质上来讲，原生 API 调用 Flutter API 也是通过 MethodChannel 来实现的，不过这里要使用的是经过封装的 MethodChannel：EventChannel。

按照上一节的实现顺序，我们先来编写原生 Android 端（服务器端）的代码逻辑，比较简单。首先需要定义一个 CHANNEL 名， Flutter 端（客户端）的 CHANNEL 名要和它一致才可以进行通信。然后定义一个 EventChannel 对象，编写逻辑，具体代码如下。

```
public class EventChannelActivity extends FlutterActivity {
```

```
    // 定义一个 CHANNEL 名，Flutter 端的 CHANNEL 名要与它一致
    public static final String STREAM = "com.flutter.eventchannel/stream";
    @BindView(R.id.toolBar)
    Toolbar toolBar;
    @Override
    protected void onCreate(Bundle savedInstanceState) {
        super.onCreate(savedInstanceState);
        setContentView(R.layout.activity_event_channel);
        ButterKnife.bind(this);
        toolBar.setTitle("EventChannel");
         // 通过 EventChannel 进行通信
        new EventChannel(getFlutterView(), STREAM).setStreamHandler(
                new EventChannel.StreamHandler() {
            @Override
            public void onListen(Object args, final EventChannel.EventSink events) {
                        // 通过监听随时随地发送消息指令给 Flutter 端
                        Log.i("info", "adding listener");
                        events.success("原生 Android 端发送过来的指令信息");
                    }

                    @Override
                    public void onCancel(Object args) {
                        Log.i("info", "cancelling listener");
                    }
                }
        );
    }
}
```

以上代码的核心是 EventChannel 里的 onListen 方法：使用 EventChannel.EventSink 主动发送消息指令给 Flutter 端。Flutter 端收到消息指令后，将调用相关 API 执行指令中的操作。

EventChannel.EventSink 中的几个核心回调方法如下。其中，success 方法用于发送成功状态的消息指令，error 方法用于发送异常情况下的消息指令。

```
// 回调方法
public interface EventSink {
    void success(Object var1);

    void error(String var1, String var2, Object var3);
```

```
  void endOfStream();
}
```

有了原生 Android 端代码，我们来看一下 Flutter 端的代码逻辑，具体如下。

```
class _MyHomePageState extends State<MyHomePage> {
  // 定义一个 CHANNEL 名，要与原生 Android 端一致
  static const EventChannel eventChannel =
      EventChannel('com.flutter.eventchannel/stream');
  StreamSubscription _streamSubscription = null;

  String _eventString = '';

  @override
  void initState() {
    super.initState();
    // 创建 EventChannel，监听数据
    _streamSubscription = eventChannel
        .receiveBroadcastStream()
        .listen(_onEvent, onError: _onError);
  }

  void _onEvent(Object event) {
    print("原生 Android 端发送过来的: $event.toString()");
    setState(() {
      _eventString = "原生 Android 端发送过来的: $event";
    });
  }

  void _onError(Object error) {
    setState(() {
      PlatformException exception = error;
      _eventString = exception?.message ?? '错误';
    });
  }
  // 停止监听并接收消息
  void _disableEvent() {
    if (_streamSubscription != null) {
      _streamSubscription.cancel();
      _streamSubscription = null;
```

```
    }
  }

  @override
  Widget build(BuildContext context) {
    return Scaffold(
      appBar: AppBar(
        title: Text(widget.title),
      ),
      body: Center(
        child: Column(
          mainAxisAlignment: MainAxisAlignment.center,
          children: <Widget>[
            RaisedButton(
              child: Text('接收原生 Android 端发送的消息$_eventString'),
              onPressed: null,
            ),
          ],
        ),
      ),
    );
  }
}
```

这样的交互通信方式可以应用在接收原生广播信息、网络状态信息、时间变化信息、电池电量信息等场景下。

15.1.4　Flutter 组件与原生控件混合使用

Flutter 组件可以和原生平台控件混合使用，但实际开发过程中并不推荐这种方式，一是比较麻烦，再有就是可能会引起其他问题。所以本节我们简单介绍 Flutter 组件和 Android 原生控件混合使用的方法，大家了解即可。

混合使用的原理同 Flutter 与原生 API 交互原理，通过 MethodChannel 即可实现，具体代码如下。

```
// 编写 Flutter 端代码逻辑
// 添加原生布局
Future<Null> _addNativeLayout() async {
  try {
```

```
    await platform.invokeMethod('addNativeLayout');
  } on PlatformException catch (e) {}
}

// 编写原生Android端代码逻辑
private static final String CHANNEL = "samples.flutter.io/battery";

    @Override
    protected void onCreate(Bundle savedInstanceState) {
        super.onCreate(savedInstanceState);
        // Flutter插件库注册
        GeneratedPluginRegistrant.registerWith(this);

        new MethodChannel(getFlutterView(), CHANNEL).setMethodCallHandler(
                new MethodChannel.MethodCallHandler() {
              @Override
              public void onMethodCall(MethodCall call, MethodChannel.Result result) {
                switch (call.method) {
                  case "addNativeLayout":
                    // 向现有Flutter布局添加布局
                    FrameLayout v = (FrameLayout) findViewById(android.R.id.content);
                    View linearLayout = new LinearLayout(MainActivity.this);
                    linearLayout.setBackgroundColor(0xff00BFFF);
                    ViewGroup.MarginLayoutParams marginLayoutParams =
                                new ViewGroup.MarginLayoutParams(600, 600);
                    ((LinearLayout) linearLayout).setGravity(Gravity.CENTER);
                    marginLayoutParams.setMargins(200, 230, 0, 0);
                    linearLayout.setLayoutParams(marginLayoutParams);
                    v.addView(linearLayout);
                    TextView textView = new TextView(MainActivity.this);
                    textView.setText("我是原生布局");
                    textView.setTextColor(Color.parseColor("#FFFFFF"));
                    textView.setGravity(Gravity.CENTER);
                    ((LinearLayout) linearLayout).addView(textView);
                  break;
                  default:
                    result.notImplemented();
                }
              }
```

```
            });
    }
```

通过以上代码可以看到，原生 Ardroid 端的核心逻辑是，通过 findViewById (android.R.id.content)方法获取 UI 的最外层父布局 FrameLayout，然后向里面动态添加 View。

15.1.5 Flutter 页面跳转到原生页面

本节我们介绍如何从 Flutter 页面跳转到原生 Android 页面，这也属于 Flutter 与原生交互的一个场景。跳转页面其实也是通过 MethodChannel 来实现的，只需定义一个跳转方法即可，具体代码如下。

```
// 编写 Flutter 端方法
// 跳转到原生页面
Future<Null> _toNativeActivity() async {
  try {
    await platform.invokeMethod('toNativeActivity');
  } on PlatformException catch (e) {}
}

// 编写原生 Android 端逻辑
public class MainActivity extends FlutterActivity {
    private static final String CHANNEL = "samples.flutter.io/battery";

    @Override
    protected void onCreate(Bundle savedInstanceState) {
        super.onCreate(savedInstanceState);
        // Flutter 插件库注册
        GeneratedPluginRegistrant.registerWith(this);

        new MethodChannel(getFlutterView(), CHANNEL).setMethodCallHandler(
                new MethodChannel.MethodCallHandler() {
             @Override
             public void onMethodCall(MethodCall call, MethodChannel.Result result) {
               switch (call.method) {
                case "toNativeActivity":
                 // 页面跳转
                 Intent intent = new Intent(MainActivity.this, NativeActivity.class);
                 startActivity(intent);
                break;
```

```
                default:
                    result.notImplemented();
                }
            }
        });
    }
}
```

通过以上代码可以看到，实现页面跳转的本质与之前讲过的 Flutter 调用原生 API 的方式一致：原生 Android 端接收到调用请求后使用 Intent 来定义跳转逻辑。

这里建议新建一个 Android Studio 项目窗口并打开 Android 项目目录（不是 Flutter 项目目录），这样就可以用 Android Studio 编写原生 Android 端代码逻辑了。

15.1.6 原生页面跳转到 Flutter 页面

上一节我们介绍了从 Flutter 页面跳转到原生页面的方法（以 Android 页面为例），接下来我们看一看如何从原生页面跳转到 Flutter 页面。从原生页面跳转到 Flutter 页面，意味着原生页面里面混合了 Flutter 页面。首先我们要使用 Android Studio 新建 Flutter Module，如图 15-3 和 15-4 所示。

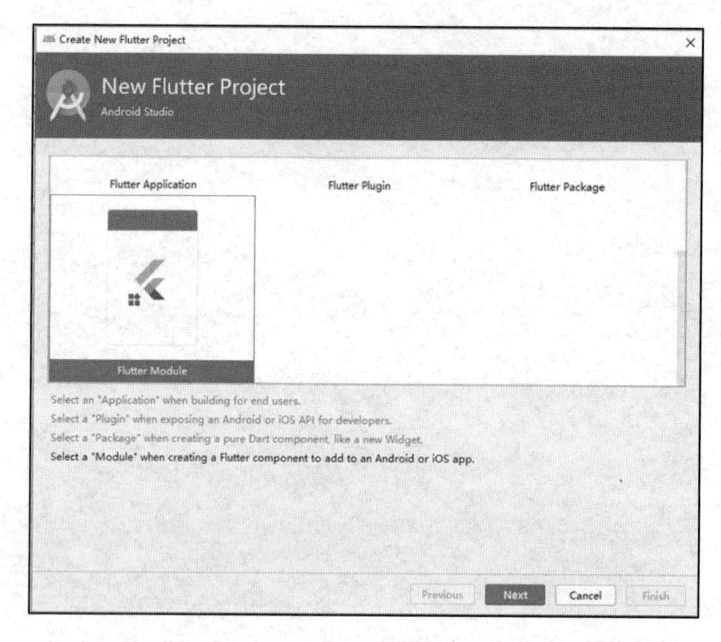

图 15-3　使用 Android Studio 新建 Flutter Module 步骤一

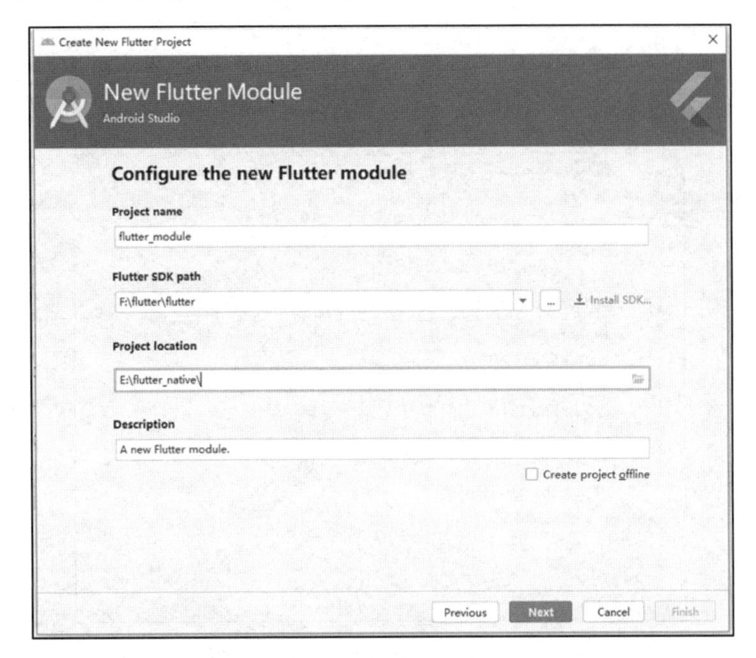

图 15-4　使用 Android Studio 新建 Flutter Module 步骤二

新建后的 Flutter Module（名为 flutter_module）结构如图 15-5 所示。

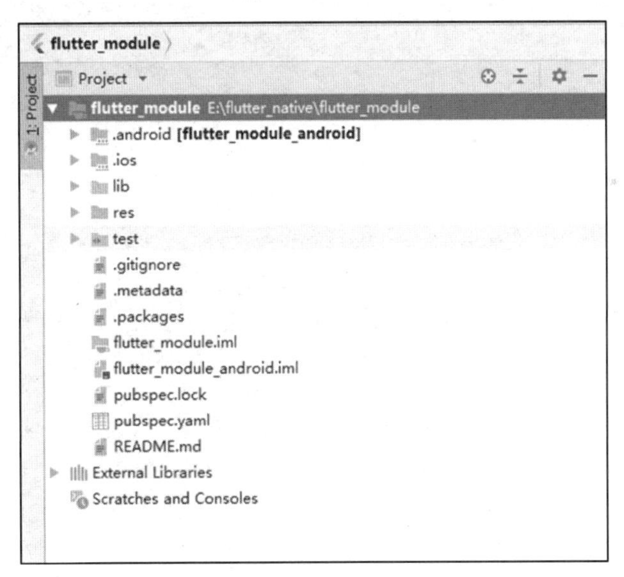

图 15-5　Flutter Module 结构

我们在新的窗口打开 Android 项目，这样方便编写 Android 代码逻辑，如图 15-6 所示。

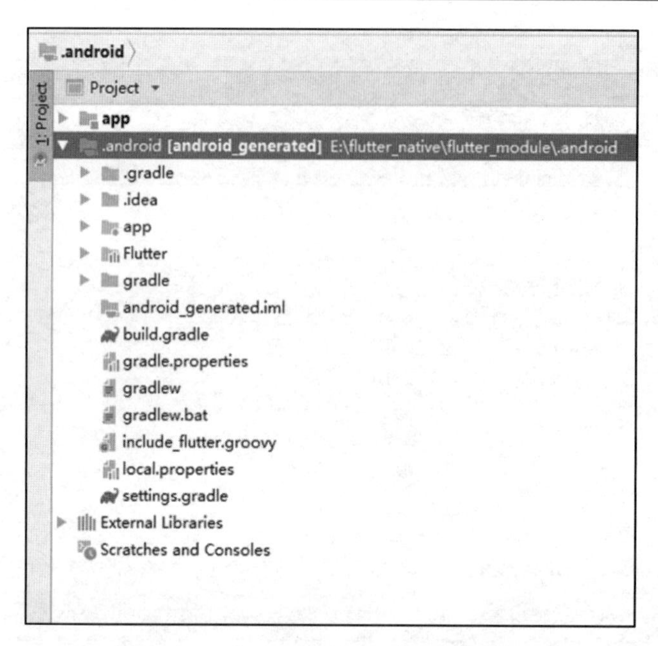

图 15-6　在新的窗口打开 Android 项目

部分原生 Android 端代码逻辑如下。

```
// 新建按钮，按下按钮跳转到 Flutter 页面
 @Override
    public void onClick(View v) {
        switch (v.getId()) {
            case R.id.fab:
                Intent intent
                        = new Intent(this, NativeFlutterActivity.class);
                startActivity(intent);
                break;
        }
    }
```

以上代码的功能是，按下按钮跳转到 Flutter 页面。接下来新建一个独立的承载 Flutter 页面的 Activity 原生页面，这里命名为 NativeFlutterActivity，代码如下。

```
package com.flutter.flutter_module.host;

import android.support.design.widget.FloatingActionButton;
import android.support.v4.app.FragmentTransaction;
import android.support.v7.app.AppCompatActivity;
```

```
import android.os.Bundle;
import android.view.View;
import android.widget.FrameLayout;

import io.flutter.facade.Flutter;
import io.flutter.view.FlutterView;

public class NativeFlutterActivity extends AppCompatActivity {
    private FrameLayout fl_container;

    @Override
    protected void onCreate(Bundle savedInstanceState) {
        super.onCreate(savedInstanceState);
        setContentView(R.layout.activity_flutter);
        fl_container = findViewById(R.id.fl_container);

        // 第一种方式：Flutter.createView，route1 为自定义的路由名称
        FlutterView flutterView = Flutter.createView(
                NativeFlutterActivity.this,
                getLifecycle(),
                "route1"
        );
        fl_container.addView(flutterView);

        // 第二种方式：Flutter.createFragment，route1 为自定义的路由名称
        // FragmentTransaction fragmentTransaction =
                        getSupportFragmentManager().beginTransaction();
        // fragmentTransaction.replace(R.id.fl_container,
                                   Flutter.createFragment("route1"));
        // fragmentTransaction.commit();

        // 为了避免跳转黑屏，可以在页面第一帧绘制出来后显示布局
        final FlutterView.FirstFrameListener[] listeners =
                                 new FlutterView.FirstFrameListener[1];
        listeners[0] = new FlutterView.FirstFrameListener() {
            @Override
            public void onFirstFrame() {
                fl_container.setVisibility(View.VISIBLE);
            }
```

```
        };
        flutterView.addFirstFrameListener(listeners[0]);
    }
}

// 布局
<?xml version="1.0" encoding="utf-8"?>
<FrameLayout xmlns:android="http://schemas.android.com/apk/res/android"
    xmlns:tools="http://schemas.android.com/tools"
    xmlns:app="http://schemas.android.com/apk/res-auto"
    android:id="@+id/fl_container"
    android:layout_width="match_parent"
    android:background="@color/white"
    android:visibility="invisible"
    android:layout_height="match_parent"
    tools:context=".NativeFlutterActivity">
</FrameLayout>

// 避免黑屏，将主题设置为透明模式，背景色改为白色
<style name="AppTheme" parent="Theme.AppCompat.Light.NoActionBar">
    <!-- Customize your theme here. -->
    <item name="android:windowBackground">@drawable/launch_background</item>
    <item name="android:windowIsTranslucent">true</item>
</style>

// 在项目清单里注册 Activity 原生页面并设置好主题
<activity
    android:name=".NativeFlutterActivity"
    android:theme="@style/AppTheme" />
```

以上为原生使用 Android 端承载调用 Flutter 端页面的功能逻辑，执行以上代码即可加载 Flutter 页面。接下来需要再编写 Flutter 端代码逻辑，如下。

```
import 'dart:ui';

import 'package:flutter/material.dart';
import 'package:flutter/services.dart';

import 'practice_two_samples.dart';
```

```
// void main() => runApp(MyApp());

void main() => runApp( _widgetForRoute(window.defaultRouteName));

// 处理路由跳转
Widget _widgetForRoute(String route) {
  switch (route) {
    case 'route1':
      return MyApp();
    case 'route2':
      return MaterialApp(
        title: 'Flutter with Native',
        theme: ThemeData(
          primarySwatch: Colors.teal,
        ),
        home: PracticeTwoSamples(),
      );
    default:
      return Center(
        child: Text('Unknown route: $route', textDirection: TextDirection.ltr),
      );
  }
}

class MyApp extends StatelessWidget {
  @override
  Widget build(BuildContext context) {
    return MaterialApp(
      title: 'Flutter with Native',
      theme: ThemeData(
        primarySwatch: Colors.teal,
      ),
      home: MyHomePage(title: 'Flutter with Native'),
    );
  }
}

class MyHomePage extends StatefulWidget {
  MyHomePage({Key key, this.title}) : super(key: key);
```

```
  final String title;

  @override
  _MyHomePageState createState() => _MyHomePageState();
}

class _MyHomePageState extends State<MyHomePage> {
  static const platform = const MethodChannel('samples.flutter.io/battery');

  // 获取电池电量信息
  String _batteryLevel = 'Unknown battery level.';
  bool _hasPermission = false;

  @override
  Widget build(BuildContext context) {
    return Scaffold(
      appBar: AppBar(
        title: Text(widget.title),
      ),
... ...
```

如果想处理 Flutter 的返回事件，就要重写 Activity 的返回事件，代码如下。

```
 @Override
    public void onBackPressed() {
        if (this.flutterView != null) {
            this.flutterView.popRoute();
        } else {
            super.onBackPressed();
        }
    }
```

以上我们实现了由 Android 端原生页面跳转到 Flutter 页面的逻辑编写，其中最值得注意的是载承 Flutter 页面的 Activity 的逻辑编写。

15.2　Flutter 插件库开发

在使用 Flutter 开发时，某些功能实现起来可能比较麻烦，或者干脆无法实现，这时我们首先应该想到搜索 Dart Pub 插件库。这是 Flutter 官方提供的针对 Flutter、Web、Dart 的开源插件仓库。本节我们将介绍 Flutter 中 Dart Pub 的使用，以及自己开发 Flutter 插件库的方法。

15.2.1　Dart Pub 的使用

Dart Pub 里面有很多官方提供的插件库可供开发者使用，还有许多其他开发者提供的开源插件库。Dart Pub 的界面如图 15-7 所示。

图 15-7　Dart Pub 界面

这里的 Flutter 插件库都是兼容 Android 和 iOS 平台的，还有很多与 Web 相关，其中使用频率较高的如图 15-8 所示。

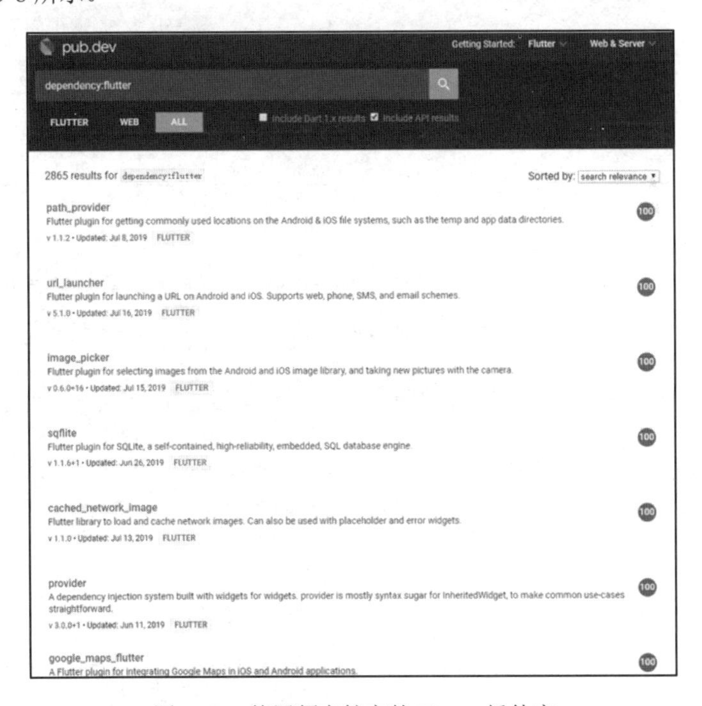

图 15-8　使用频率较高的 Flutter 插件库

这里以 path_provider 插件库为例进行简单说明。单击 path_provider 链接进入该插件库详情描述页，如图 15-9 所示。该页面展示了说明文档、更新日志、示例代码、安装步骤、版本列表、关于页面、插件库官方首页、GitHub 地址、相关问题、API 文档等信息。

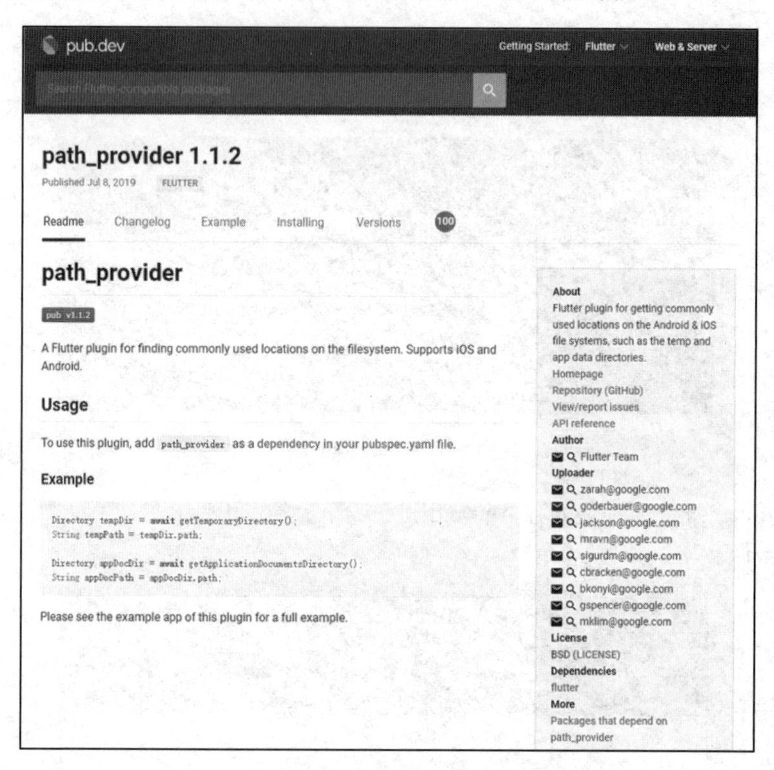

图 15-9　path_provider 插件库详情描述页

path_provider 主要用于为 Flutter 提供访问 Android 和 iOS 文件系统目录的功能。我们需要选择 Installing，将安装依赖库的信息复制到 pubspec.yaml 中。

```
dependencies:
  path_provider: ^1.1.2
```

接下来可以在 Readme 或者 Example 里看到 path_provider 的使用方法和说明，如果没有提供这些信息，可以进入 path_provider 官方 GitHub 页面下载和查看示例代码。

15.2.2　Flutter Package 开发

除了可以使用 Dart Pub 中提供的插件库，我们还可以自己编写插件库。不过插件库必须能够兼容 Android 和 iOS 平台，也就是说在开发时要分别编写 Android 和 iOS 原生代码，并提供给 Flutter 进行调用。开发完成，我们也可以将插件库提交到 Dart Pub 上供他人使用。

本节就来介绍如何编写 Flutter 插件库并提交到 Dart Pub 上。

Dart Pub 上的插件库按照实现方式不同主要分为两种：一种是包（Flutter Package），即完全通过 Dart 编写的 API 插件库；另一种是插件（Flutter Plugin），即通过编写 Android、iOS 原生代码然后使用 MethodChannel 来调用的插件库。

这两种插件库各有各的特点和优势，在创建项目时，Android Studio 也提供了两种创建插件库的方式，如图 15-10 中框起来的部分。我们需要根据实际需求选择其中一种方式。

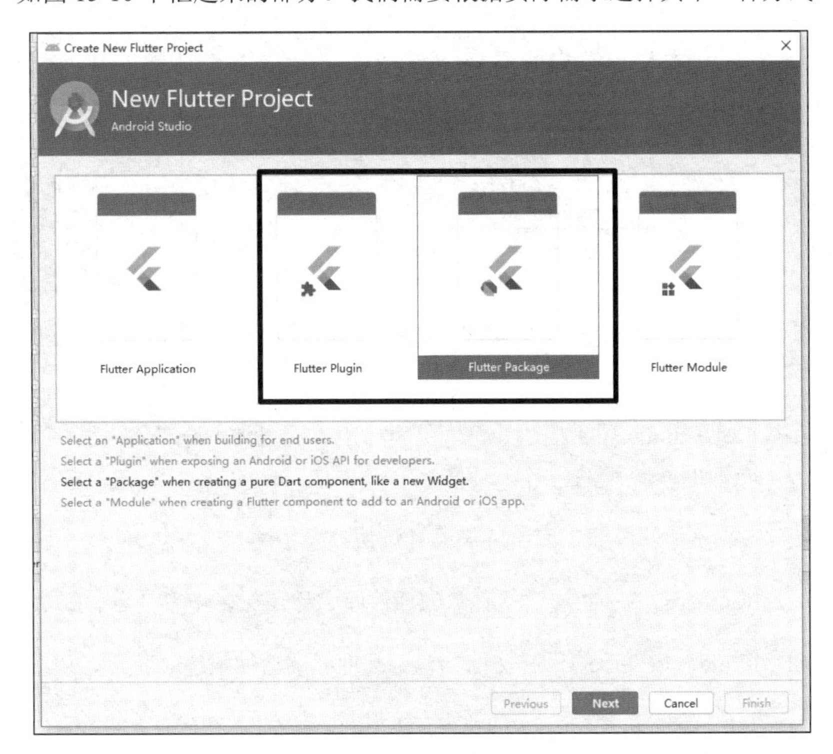

图 15-10 创建插件库的方式

根据前面的介绍，如果插件库可以完全由 Dart 编写实现，就选择 Flutter Package 创建方式；如果插件库必须借助 Android、iOS 原生代码才可以实现，就需要选择 Flutter Plugin 创建方式。

先来看一下如何创建 Flutter Package 插件库。除了可以在 Android Studio 界面进行操作，还可以通过命令行方式创建，如下。

```
$ flutter create --template=package hello
```

这里的 hello 为插件库的英文名称，新建插件库的结构如图 15-11 所示。

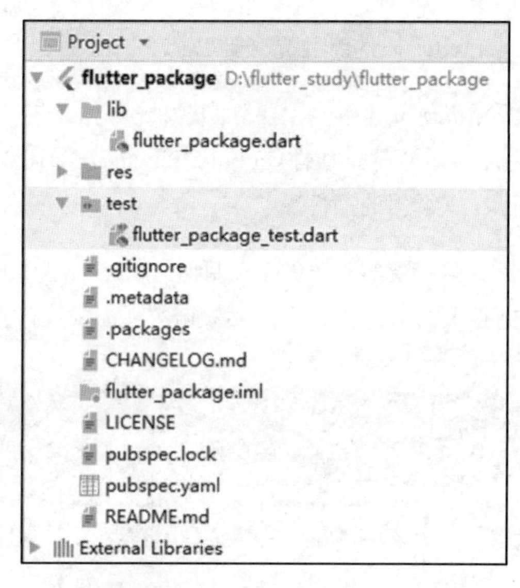

图 15-11　新建插件库的结构

插件库逻辑需写在 lib 目录下，完全是通过 Dart 实现的。test 目录下面则存放单元测试文件。新建插件库项目里主要包含以下内容。

- CHANGELOG.md：用于描述插件库更新日志。提交到 Dart Pub 上就是插件库主页的 Changelog 项目里显示的内容。
- LICENSE：添加开源插件库所遵循的开源协议及描述。
- pubspec.yaml：插件库的描述信息，如插件库名称信息、作者信息、版本信息等。
- README.md：插件库的说明文档，包含插件库的功能描述、特点、用法示例等。

我们可以大致看一下 pubspec.yaml 的内容，如下。

```
// 插件库名称
name: flutter_package
// 插件库描述
description: A new Flutter package.
// 插件库版本
version: 0.0.1
// 作者信息，名字 <邮箱>
author: Tan Dong <tandongjay@qq.com>
```

```
// 插件库主页地址
homepage: https://github.con/tandong/flutter_package
// 插件库运行的SDK 版本环境
environment:
  sdk: ">=2.1.0 <3.0.0"
// 所需依赖
dependencies:
  flutter:
    sdk: flutter

dev_dependencies:
  flutter_test:
    sdk: flutter
```

假如插件库名称为 flutter_package，则默认会在 lib 目录下新建一个 flutter_package.dart 文件，该文件的内容如下。

```
library flutter_package;

/// 计算功能核心类
class Calculator {
  /// 加 1 操作
  int addOne(int value) => value + 1;
}
```

当插件库项目结构中具备了上述文件和目录，并且在 lib 目录下完成了 Dart 功能代码编写后，就可以发布到 Dart Pub 上了。

首先运行如下命令，检查插件库发布所需文件及结构是否齐全且设置正确。

```
$ flutter packages pub publish --dry-run
```

如果检查通过，输入如下命令即可完成发布。

```
$ flutter packages pub publish
```

命令窗口发布过程如图 15-12 所示。

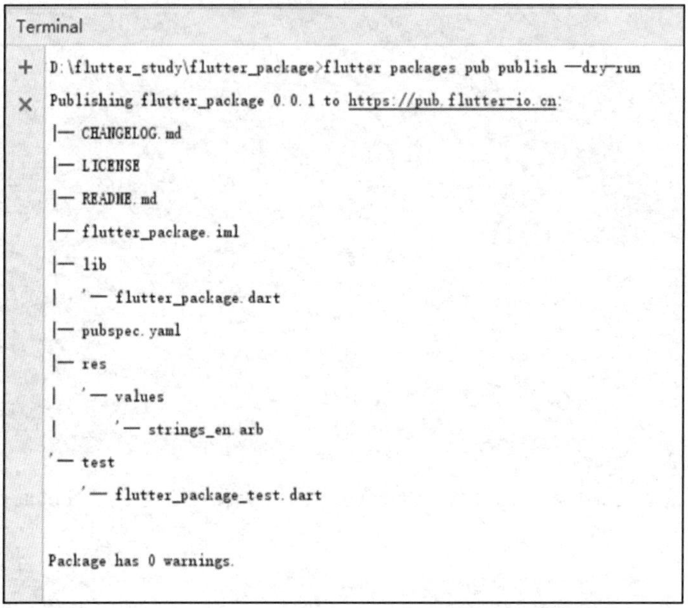

```
Terminal
D:\flutter_study\flutter_package>flutter packages pub publish --dry-run
Publishing flutter_package 0.0.1 to https://pub.flutter-io.cn:
|-- CHANGELOG.md
|-- LICENSE
|-- README.md
|-- flutter_package.iml
|-- lib
|   '-- flutter_package.dart
|-- pubspec.yaml
|-- res
|   '-- values
|       '-- strings_en.arb
'-- test
    '-- flutter_package_test.dart

Package has 0 warnings.
```

图 15-12　命令窗口发布过程

如果遇到第三方插件库的依赖库版本与项目中引用的插件库依赖库版本不一致的问题，Flutter 官方建议在插件库所用的依赖库版本声明中声明依赖库的版本号范围，而不指定某个特定版本号。例如 flutter_package 插件库的 pubspec.yaml 中声明了如下依赖库版本，表明版本为 0.4.2 及以上，这样的声明是推荐的。

```
dependencies:
  url_launcher: ^0.4.2
```

还有一种方式可以解决上述问题，即强制使用 dependency_overrides 来指定依赖库的版本，如下。

```
dependencies:
  some_package:
  other_package:
dependency_overrides:
  url_launcher: '0.4.3'
```

以上介绍了 Flutter Package 插件库的开发与发布，以及常见问题的解决方案。大家可以尝试开发一个自己的插件库，并发布到 Dart Pub 上与他人共享。

15.2.3 Flutter Plugin 开发

本节将介绍另一种插件库 Flutter Plugin 的开发过程。开发 Flutter Plugin 插件库时需要编写 Android 和 iOS 原生代码，再通过交互调用实现特定功能。

如同前面介绍的，创建时要在 Android Studio 里选择 Flutter Plugin，也可以使用如下的命令行方式创建。这里建议使用 Android Studio 创建。

```
$ flutter create --org com.example --template=plugin hello
```

默认的开发语言是 Java 和 Objective-C，我们也可以在命令行中指定开发语言，如指定 Kotlin 或 Swift。

```
$ flutter create --template=plugin -i swift -a kotlin hello
```

创建后的 Flutter Plugin 项目（名为 flutter_plugin）的结构如图 15-13 所示。

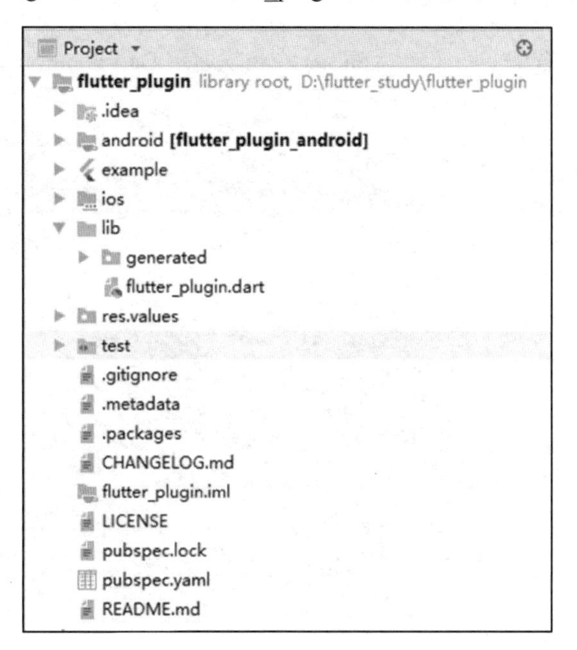

图 15-13 Flutter Plugin 项目的结构

通过图 15-13 可以看到，Flutter Plugin 项目结构和默认的 Flutter 项目结构很像，其中各个目录和文件的具体含义大致如下。

- android：用于存放编写好的 Android 原生代码。
- ios：用于存放编写好的 iOS 原生代码。

- example：用于存放编写好的插件库使用示例。
- lib：用于存放编写插件的 Flutter 代码文件，内含调用封装好的方法。
- test：用于存放单元测试目录。
- CHANGELOG.md：插件库更新日志的描述文件。
- LICENSE：插件库所遵循的开源协议和说明。
- pubspec.yaml：插件库的配置描述信息，如名称、描述、作者、主页地址、版本、运行环境等。
- README.md：插件库说明文档。

Flutter Plugin 插件库的发布步骤同 Flutter Package 的发布步骤一致，这里不再重复讲解。

扩展一下，我们来看一下如何使用 Flutter 中未发布到 Dart Pub 的本地插件库。本地插件库的引入方式如下。

```
dependencies:
  plugin1:
    path: ../plugin1/
```

GitHub 上托管的插件库的引入方式如下。

```
dependencies:
  package1:
    git:
      url: git://github.com/flutter/packages.git
      path: packages/package2
```

通过以上方式我们可以使用自己编写的本地插件库，而无须发布到 Dart Pub 上再引用。其实插件库开发和 Flutter 与原生 API 交互有很多的关联，希望大家边阅读边思考，多多实践。

第16章

Flutter 调试与应用打包发布

无论使用什么开发语言进行开发，一般都会用到调试与单元测试功能。Flutter 也不例外，它也支持调试与单元测试。本章将介绍 Flutter 调试、单元测试、辅助工具的使用，同时介绍 Flutter 应用打包与发布。

16.1 调试与单元测试

应用的调试与单元测试是开发中必不可少的步骤，Flutter 也提供了很多实用的调试方法，如日志调试、断点调试、真机调试、可视化调试等。Flutter 单元测试与其他平台的单元测试大同小异，具体方式均为：自动生成一个 test 测试文件夹，在里面编写测试文件和测试用例。本节我们就来具体说说 Flutter 调试与单元测试，并拓展介绍一些辅助工具的使用。

16.1.1 调试

调试可用于查找和分析问题，验证某些输入、输出数据是否正确。Flutter 的调试功能非常强大，我们先来看一个简单的例子，通过输出控制台日志来查看调试程序，如下。

```
// 用 print()方法来输出控制台日志
print(object)
// 可以这样使用
int a = 6;
double b = 3.18;
print('$a ,$b');
// print()方法取值是通过$符号实现的
// 还可以通过 debugPrint()来输出日志，参数只能是 String 类型的
debugPrint(String);
```

上述调试方式称为日志调试，简单直接，使用频率非常高。当一次输出太多日志时，Android 有时会丢弃一些日志行。为了避免这种情况发生，可以使用 Flutter foundation 库中的 debugPrint()方法。这是一个封装过的 print()方法，可以避免丢弃日志行。我们可以在 Android Studio 的命令行窗口通过 flutter logs 命令来过滤并查看输出的日志。

接下来我们介绍断点调试，这也是比较常用的调试方式。首先以 Android Studio 为例，只要在需要添加断点的语句左侧单击鼠标左键，就能添加一个红色的断点，如图 16-1 所示。

```
29
30    void _incrementCounter() {
31      setState(() {
32        _counter++;
33      });
34      print(_counter);
35    }
36
```

图 16-1　在 Android Studio 中添加断点

接下来以 Debug 模式运行应用程序，如图 16-2 所示。

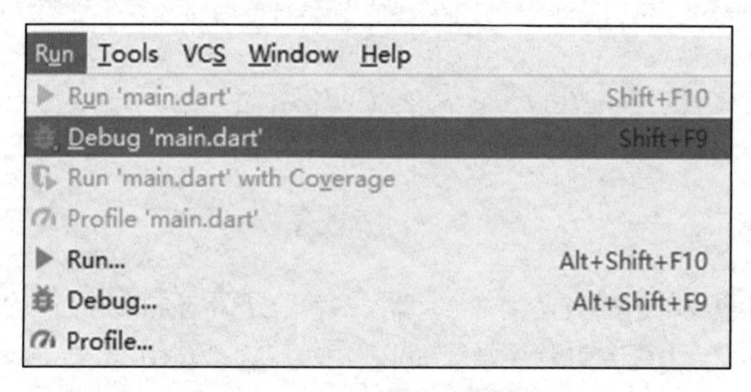

图 16-2　以 Debug 模式运行应用程序

在图 16-3 中，我们在_incrementCounter 方法中设置了两个断点，当执行_incrementCounter 方法时程序会自动执行到断点处暂停，查看断点调试。在调试控制台里可以看到断点处的信息和属性值，包括之前设置了断点的_counter 变量。另外，当前断点所在行会高亮标识，当鼠标移动到断点所在行时会出现 Run to Cusor 工具提示，用鼠标左键单击调试窗口中的 Run to Cusor(Alt+F9)选项可以跳到下一个断点。

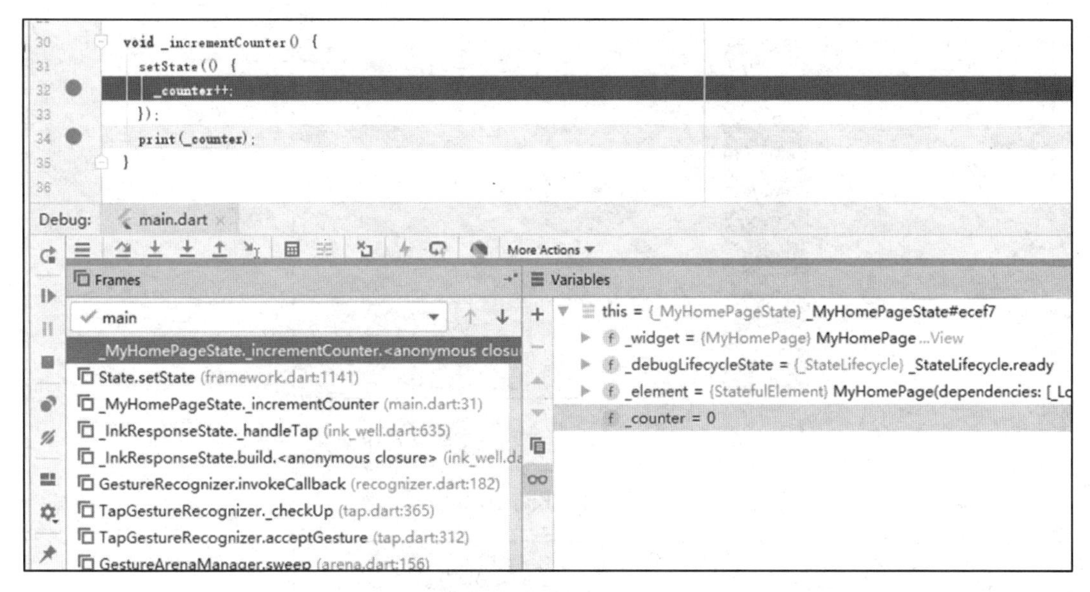

图 16-3　查看断点调试

VSCode 的断点调试也是一样的。如果使用 VSCode 开发工具，只需单击要进行断点调试的语句所在行的左侧边栏，就会添加一个红色的断点，如图 16-4 所示。

```
75
76    void _changeTextContent() {
77      setState(() {
78        //setState里用于刷新UI和绑定数据
79        title = change ? "这个图片很好看,
80        change = !change;
81      });
82      gotoPage();
83    }
84
85    void gotoPage() {
86      Navigator.push(context, MaterialPa
87        return EventBusSamples();
88      })); // MaterialPageRoute
89    }
90
```

图 16-4　在 VSCode 中添加断点

接着同样以 Debug 模式运行应用程序，如图 16-5 所示。

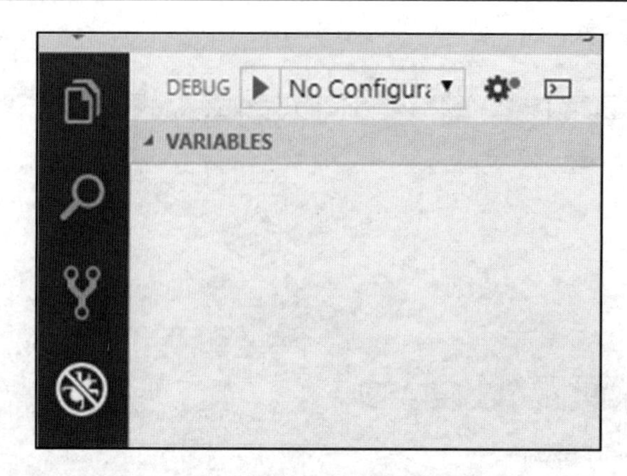

图 16-5　以 Debug 模式运行应用程序

这时编辑器页面顶部会悬浮调试操作栏，如图 16-6 所示。

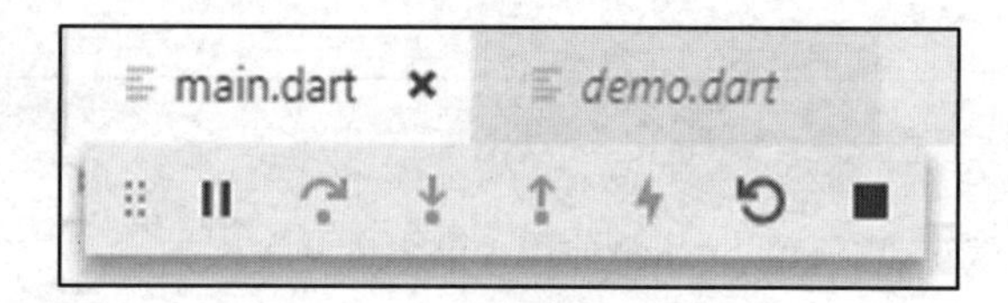

图 16-6　调试操作栏

在 Debug 窗口可以针对相关断点查看其变量属性值，如图 16-7 所示。

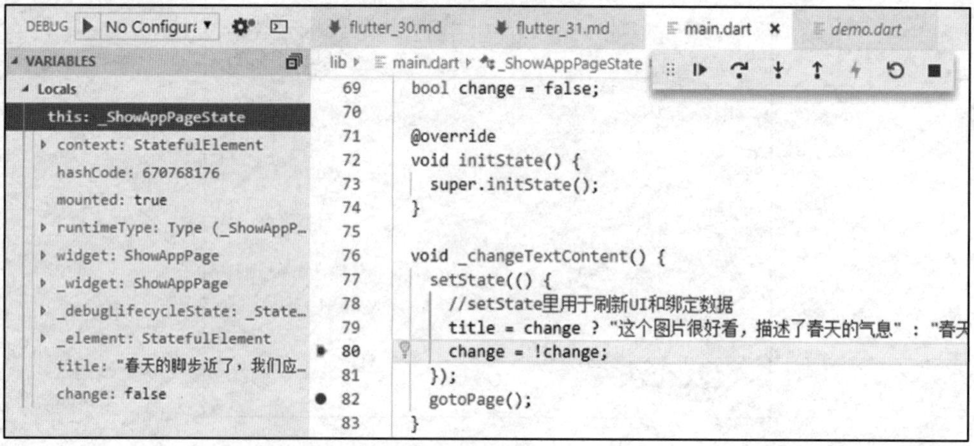

图 16-7　查看断点的变量属性值

除了上面介绍的调试方式，Flutter 还支持通过命令行方式进行调试。例如，可以通过 Dart 分析器分析代码并发现其中的问题。在项目所在根目录中输入命令 flutter analyze 即可进行代码分析与调试，如图 16-8 所示。

```
D:\flutter_app\flutter_app>flutter analyze
Analyzing flutter_app...
No issues found! (ran in 3.1s)

D:\flutter_app\flutter_app>
```

图 16-8　代码分析与调试

另外，我们可以在代码里植入一些调试专用 API 语句来实现调试功能。例如，debugger()调试 API 的构造方法如下。

```
external bool debugger({bool when: true, String message});
```

具体使用示例如下。

```
import 'dart:developer';
... ...
  void _incrementCounter() {
    setState(() {
      _counter++;
    });
    print(_counter);
    // 满足某个条件时进行调试
    debugger(when: _counter > 5.0);
  }
```

我们还可以设置 assert 判断变量、常量或表达式是否等于某个值，具体使用示例如下。

```
assert(_counter==2);
```

如果想查看 Flutter 页面组件的树形结构，可以使用 debugDumpApp()方法，图 16-9 展示了某页面组件的树形结构。通过 debugDumpApp()方法可以详细看到整个页面涉及的组件内容，需要注意的是，debugDumpApp()方法要在调用 runApp()方法之后使用。

```
I/flutter ( 7751): 1
I/flutter ( 7751): WidgetsFlutterBinding - CHECKED MODE
I/flutter ( 7751): [root](renderObject: RenderView#c6e24)
I/flutter ( 7751):  └MyApp
I/flutter ( 7751):   └MaterialApp(state: _MaterialAppState#2310f)
I/flutter ( 7751):    └ScrollConfiguration(behavior: _MaterialScrollBehavior)
I/flutter ( 7751):     └WidgetsApp-[GlobalObjectKey _MaterialAppState#2310f](state: _WidgetsAppState#910e3)
I/flutter ( 7751):      └DefaultFocusTraversal
I/flutter ( 7751):       └MediaQuery(MediaQueryData(size: Size(411.4, 774.9), devicePixelRatio: 3.5, textScaleFactor: 1.0, platformBrightness: Brightness.
I/flutter ( 7751):        └Localizations(locale: en_US, delegates: [DefaultMaterialLocalizations.delegate(en_US), DefaultCupertinoLocalizations.delegate(e
I/flutter ( 7751):         └Semantics(container: false, properties: SemanticsProperties, label: null, value: null, hint: null, textDirection: ltr, hintOve
I/flutter ( 7751):          └_LocalizationsScope-[GlobalKey#3de5a]
I/flutter ( 7751):           └Directionality(textDirection: ltr)
I/flutter ( 7751):            └Title(title: "Flutter Demo", color: MaterialColor(primary value: Color(0xff2196f3)))
I/flutter ( 7751):             └CheckedModeBanner("DEBUG")
I/flutter ( 7751):              └Banner("DEBUG", textDirection: ltr, location: topEnd, Color(0xa0b71c1c), text inherit: true, text color: Color(0xffffffff
I/flutter ( 7751):               └CustomPaint(renderObject: RenderCustomPaint#3600c)
I/flutter ( 7751):                └DefaultTextStyle(debugLabel: fallback style; consider putting your text in a Material, inherit: true, color: Color(0xd0
I/flutter ( 7751):                 └Builder(dependencies: [MediaQuery])
I/flutter ( 7751):                  └AnimatedTheme(duration: 200ms, state: _AnimatedThemeState#4a07d(ticker inactive, ThemeDataTween(ThemeData#2099e(butto
I/flutter ( 7751):                   └Theme(ThemeData#2099e(buttonTheme: ButtonThemeData#c9338(buttonColor: Color(0xffe0e0e0), focusColor: Color(0x1f00000
I/flutter ( 7751):                    └_InheritedTheme
I/flutter ( 7751):                     └CupertinoTheme
I/flutter ( 7751):                      └_InheritedCupertinoTheme
I/flutter ( 7751):                       └IconTheme(IconThemeData#79ebd(color: MaterialColor(primary value: Color(0xff2196f3))))
I/flutter ( 7751):                        └IconTheme(IconThemeData#2abdc(color: Color(0xdd000000)))
I/flutter ( 7751):                         └Navigator-[GlobalObjectKey<NavigatorState> _WidgetsAppState#910e3](state: NavigatorState#c7a10(tickers: track
I/flutter ( 7751):                          └Listener(listeners: [down, up, cancel], behavior: deferToChild, renderObject: RenderPointerListener#ce18f)
I/flutter ( 7751):                           └AbsorbPointer(absorbing: false, renderObject: RenderAbsorbPointer#512ce)
I/flutter ( 7751):                            └FocusScope(AUTOFOCUS, node: FocusScopeNode#52ba7, state: _FocusScopeState#978b2)
I/flutter ( 7751):                             └Semantics(container: false, properties: SemanticsProperties, label: null, value: null, hint: null, hintOve
I/flutter ( 7751):                              └_FocusMarker
I/flutter ( 7751):                               └Overlay-[LabeledGlobalKey<OverlayState>#c403d](state: OverlayState#1c986(entries: [OverlayEntry#a62c5(op
I/flutter ( 7751):                                └_Theatre(renderObject: _RenderTheatre#57dfd)
```

图 16-9　页面组件的树形结构

如果觉得通过 debugDumpApp()方法输出的页面组件树形结构不够详细，还可以使用 Flutter 提供的 debugDumpRenderTree()方法获取更加清晰和详细的页面组件树形结构。

使用 debugDumpRenderTree()方法时需要导入包，命令如下。

```
import'package:flutter/rendering.dart';
```

通过 debugDumpRenderTree()方法获取的页面组件树形结构如图 16-10 所示。

```
I/flutter ( 7751): 1
I/flutter ( 7751): TransformLayer#e34b0
I/flutter ( 7751):  │ owner: RenderView#cdadf
I/flutter ( 7751):  │ creator: [root]
I/flutter ( 7751):  │ offset: Offset(0.0, 0.0)
I/flutter ( 7751):  │ transform:
I/flutter ( 7751):  │   [0] 3.5,0.0,0.0,0.0
I/flutter ( 7751):  │   [1] 0.0,3.5,0.0,0.0
I/flutter ( 7751):  │   [2] 0.0,0.0,1.0,0.0
I/flutter ( 7751):  │   [3] 0.0,0.0,0.0,1.0
I/flutter ( 7751):  │
I/flutter ( 7751):  ├─child 1: OffsetLayer#8e308
I/flutter ( 7751):  │ │ creator: RepaintBoundary ← _FocusMarker ← Semantics ← FocusScope
I/flutter ( 7751):  │ │   ← PageStorage ← Offstage ← _ModalScopeStatus ←
I/flutter ( 7751):  │ │   _ModalScope<dynamic>-[LabeledGlobalKey<_ModalScopeState<dynamic>>#70d41]
I/flutter ( 7751):  │ │   ← _OverlayEntry-[LabeledGlobalKey<_OverlayEntryState>#1bb5c] ←
I/flutter ( 7751):  │ │   Stack ← _Theatre ←
I/flutter ( 7751):  │ │   Overlay-[LabeledGlobalKey<OverlayState>#c3bbd] ← ···
I/flutter ( 7751):  │ │ offset: Offset(0.0, 0.0)
I/flutter ( 7751):  │ │
I/flutter ( 7751):  │ └─child 1: OffsetLayer#f308f
I/flutter ( 7751):  │   │ creator: RepaintBoundary-[GlobalKey#3acc6] ← IgnorePointer ←
I/flutter ( 7751):  │   │   FadeTransition ← FractionalTranslation ← SlideTransition ←
I/flutter ( 7751):  │   │   _FadeUpwardsPageTransition ← AnimatedBuilder ← RepaintBoundary
I/flutter ( 7751):  │   │   ← _FocusMarker ← Semantics ← FocusScope ← PageStorage ← ···
I/flutter ( 7751):  │   │ offset: Offset(0.0, 0.0)
I/flutter ( 7751):  │   │
I/flutter ( 7751):  │   └─child 1: PhysicalModelLayer#7e02d
I/flutter ( 7751):  │     │ creator: PhysicalModel ← AnimatedPhysicalModel ← Material ←
I/flutter ( 7751):  │     │   PrimaryScrollController ← _ScaffoldScope ← Scaffold ←
I/flutter ( 7751):  │     │   MyHomePage ← Semantics ← Builder ←
I/flutter ( 7751):  │     │   RepaintBoundary-[GlobalKey#3acc6] ← IgnorePointer ←
I/flutter ( 7751):  │     │   FadeTransition ← ···
I/flutter ( 7751):  │     │ elevation: 0.0
```

图 16-10　通过 debugDumpRenderTree()方法获取的页面组件树形结构

类似的调试 API 还有 debugDumpLayerTree()方法，主要用于获取层级调试信息，效果如图 16-11 所示。

```
I/flutter ( 7751): 1
I/flutter ( 7751): TransformLayer#e34b0
I/flutter ( 7751):  │ owner: RenderView#cdadf
I/flutter ( 7751):  │ creator: [root]
I/flutter ( 7751):  │ offset: Offset(0.0, 0.0)
I/flutter ( 7751):  │ transform:
I/flutter ( 7751):  │   [0] 3.5,0.0,0.0,0.0
I/flutter ( 7751):  │   [1] 0.0,3.5,0.0,0.0
I/flutter ( 7751):  │   [2] 0.0,0.0,1.0,0.0
I/flutter ( 7751):  │   [3] 0.0,0.0,0.0,1.0
I/flutter ( 7751):  │
I/flutter ( 7751):  ├─child 1: OffsetLayer#8e308
I/flutter ( 7751):  │ │ creator: RepaintBoundary ← _FocusMarker ← Semantics ← FocusScope
I/flutter ( 7751):  │ │   ← PageStorage ← Offstage ← _ModalScopeStatus ←
I/flutter ( 7751):  │ │   _ModalScope<dynamic>-[LabeledGlobalKey<_ModalScopeState<dynamic>>#70d41]
I/flutter ( 7751):  │ │   ← _OverlayEntry-[LabeledGlobalKey<_OverlayEntryState>#1bb5c] ←
I/flutter ( 7751):  │ │   Stack ← _Theatre ←
I/flutter ( 7751):  │ │   Overlay-[LabeledGlobalKey<OverlayState>#c3bbd] ← ···
I/flutter ( 7751):  │ │ offset: Offset(0.0, 0.0)
I/flutter ( 7751):  │ │
I/flutter ( 7751):  │ └─child 1: OffsetLayer#f308f
I/flutter ( 7751):  │   │ creator: RepaintBoundary-[GlobalKey#3acc6] ← IgnorePointer ←
I/flutter ( 7751):  │   │   FadeTransition ← FractionalTranslation ← SlideTransition ←
I/flutter ( 7751):  │   │   _FadeUpwardsPageTransition ← AnimatedBuilder ← RepaintBoundary
I/flutter ( 7751):  │   │   ← _FocusMarker ← Semantics ← FocusScope ← PageStorage ← ···
I/flutter ( 7751):  │   │ offset: Offset(0.0, 0.0)
I/flutter ( 7751):  │   │
I/flutter ( 7751):  │   └─child 1: PhysicalModelLayer#7e02d
I/flutter ( 7751):  │     │ creator: PhysicalModel ← AnimatedPhysicalModel ← Material ←
I/flutter ( 7751):  │     │   PrimaryScrollController ← _ScaffoldScope ← Scaffold ←
I/flutter ( 7751):  │     │   MyHomePage ← Semantics ← Builder ←
I/flutter ( 7751):  │     │   RepaintBoundary-[GlobalKey#3acc6] ← IgnorePointer ←
I/flutter ( 7751):  │     │   FadeTransition ← ···
I/flutter ( 7751):  │     │ elevation: 0.0
```

图 16-11 通过 debugDumpLayerTree()方法获取的层级调试信息

我们还可以调用 debugDumpSemanticsTree()方法获取语义树，其构造方法如下。

```
debugDumpSemanticsTree(DebugSemanticsDumpOrder.traversalOrder);
```

使用时，需要开启 showSemanticsDebugger 配置属性，代码如下。

```
class MyApp extends StatelessWidget {
  @override
  Widget build(BuildContext context) {
    return MaterialApp(
      title: 'Flutter Demo',
      theme: ThemeData(
        primarySwatch: Colors.blue,
```

```
    ),
    // 开启 showSemanticsDebugger 配置属性
    showSemanticsDebugger: true,
    home: MyHomePage(title: 'Flutter Demo Home Page'),
  );
 }
}
```

通过 debugDumpSemanticsTree()方法获取的语义树如图 16-12 所示。

```
I/flutter ( 7751): 1
I/flutter ( 7751): SemanticsNode#0
I/flutter ( 7751):  | Rect.fromLTRB(0.0, 0.0, 1440.0, 2712.0)
I/flutter ( 7751):  |
I/flutter ( 7751):  └─SemanticsNode#1
I/flutter ( 7751):    | Rect.fromLTRB(0.0, 0.0, 411.4, 774.9) scaled by 3.5x
I/flutter ( 7751):    | textDirection: ltr
I/flutter ( 7751):    |
I/flutter ( 7751):    └─SemanticsNode#2
I/flutter ( 7751):      | Rect.fromLTRB(0.0, 0.0, 411.4, 774.9)
I/flutter ( 7751):      | flags: scopesRoute
I/flutter ( 7751):      |
I/flutter ( 7751):      ├─SemanticsNode#5
I/flutter ( 7751):      | | Rect.fromLTRB(0.0, 0.0, 411.4, 80.0)
I/flutter ( 7751):      | | thickness: 4.0
I/flutter ( 7751):      | |
I/flutter ( 7751):      | └─SemanticsNode#6
I/flutter ( 7751):      |     Rect.fromLTRB(16.0, 40.5, 242.0, 63.5)
I/flutter ( 7751):      |     flags: isHeader, namesRoute
I/flutter ( 7751):      |     label: "Flutter Demo Home Page"
I/flutter ( 7751):      |     textDirection: ltr
I/flutter ( 7751):      |     elevation: 4.0
I/flutter ( 7751):      |
I/flutter ( 7751):      ├─SemanticsNode#3
I/flutter ( 7751):      |   Rect.fromLTRB(65.7, 399.4, 345.7, 415.4)
I/flutter ( 7751):      |   label: "You have pushed the button this many times:"
I/flutter ( 7751):      |   textDirection: ltr
I/flutter ( 7751):      |
```

图 16-12　通过 debugDumpSemanticsTree()方法获取的语义树

接下来我们来看 Flutter 中的另一种调试方式：可视化调试。可视化调试使用简单，具体操作时，可以通过设置 debugPaintSizeEnabled 为 true 来进行可视化调试布局，代码如下。

```
... ...
 @override
  Widget build(BuildContext context) {
```

```
    debugPaintSizeEnabled = true;
    return Scaffold(
      appBar: AppBar(
        title: Text(widget.title),
      ),
      body: Center(
... ...
```

开启可视化调试的界面如图 16-13 所示。

图 16-13　开启可视化调试的界面

开启可视化调试后，Flutter 会自动绘制基线、辅助线、箭头、颜色等信息来帮助我们分析和绘制布局。可视化调试中的其他的调试命令还有 debugPrintBeginFrameBanner、debugPrintEndFrameBanner、debugPrintScheduleFrameStacks，用法类似于 debugPaintSizeEnabled，示例如下。

```
@override
 Widget build(BuildContext context) {
   debugPrintBeginFrameBanner = true;
```

```
    debugPrintEndFrameBanner = true;
    debugPrintScheduleFrameStacks = true;
    return Scaffold(
      appBar: AppBar(
        title: Text(widget.title),
      ),
      body: Center(
```

接下来我们介绍 Flutter 动画调试。当制作了一个动画后，可能会因为动画运行太快而无法观察其细节，此时我们可以通过调试动画来减慢它的速度，进行细节优化。进行 Flutter 动画调试时，只需将 timeDilation 变量（在 scheduler 库中）设置为大于 1.0 的数字即可（如 60.0），并且我们只需在应用程序启动时设置一次即可，代码如下。

```
class MyApp extends StatelessWidget {

  @override
  Widget build(BuildContext context) {
    timeDilation = 60.0;
    return MaterialApp(
      title: 'Flutter Demo',
      theme: ThemeData(
        primarySwatch: Colors.blue,
      ),
      showSemanticsDebugger: false,
      home: MyHomePage(title: 'Flutter Demo Home Page'),
    );
  }
}
```

再来看一下 Flutter 性能调试。有时应用程序会被要求重新布局或重新绘制，要想找到原因，可以分别设置 debugPrintMarkNeedsLayoutStacks 和 debugPrintMarkNeedsPaintStacks 标志。每当应用程序被要求重新布局或重新绘制时，这些标志会将堆栈跟踪记录输出至控制台。可以使用 services 库中的 debugPrintStack()方法按需打印堆栈跟踪记录，代码如下。

```
class MyApp extends StatelessWidget {
  @override
  Widget build(BuildContext context) {

    debugPrintMarkNeedsLayoutStacks = true;
    debugPrintMarkNeedsPaintStacks = true;
```

```
    debugPrintStack();

    return MaterialApp(
      title: 'Flutter Demo',
      theme: ThemeData(
        primarySwatch: Colors.blue,
      ),
      showSemanticsDebugger: false,
      home: MyHomePage(title: 'Flutter Demo Home Page'),
    );
  }
}
```

如果想查看与应用启动时间相关的信息，首先需通过以下命令运行一个 Flutter 程序。

```
$ flutter run --trace-startup --profile
```

在 Flutter 项目根目录的 build 目录下有一个 start_up_info.json 文件，各个关键阶段的启动时间信息都可以在这个文件中看到，如下。

```
// 进入 Flutter 引擎
"engineEnterTimestampMicros": 96025565262,
// 展示应用第一帧
"timeToFirstFrameMicros": 2171978,
// 开始进行 Flutter 框架初始化
"timeToFrameworkInitMicros": 514585,
// 完成 Flutter 框架初始化
"timeAfterFrameworkInitMicros": 1657393
```

如果想跟踪代码性能，可以使用 dart:developer 的 Timeline API 测试代码块，示例如下。

```
Timeline.startSync('interesting function');
// 中间执行的业务逻辑
… …
Timeline.finishSync();
```

这样我们就可以获取某个方法或代码块所执行的时间信息。

如果想直观查看应用程序的实时性能，可以开启 showPerformanceOverlay 属性，代码如下。

```
class MyApp extends StatelessWidget {
  @override
  Widget build(BuildContext context) {
    return MaterialApp(
```

```
      title: 'Flutter Demo',
      theme: ThemeData(
        primarySwatch: Colors.blue,
      ),
      showPerformanceOverlay: true,
      home: MyHomePage(title: 'Flutter Demo Home Page'),
    );
  }
}
```

开启 showPerformanceOverlay 属性后，应用界面上会显示应用程序性能图，具体为两个图表：第一个反映 GPU 线程花费时间，第二个反映 CPU 线程花费时间，如图 16-14 所示。

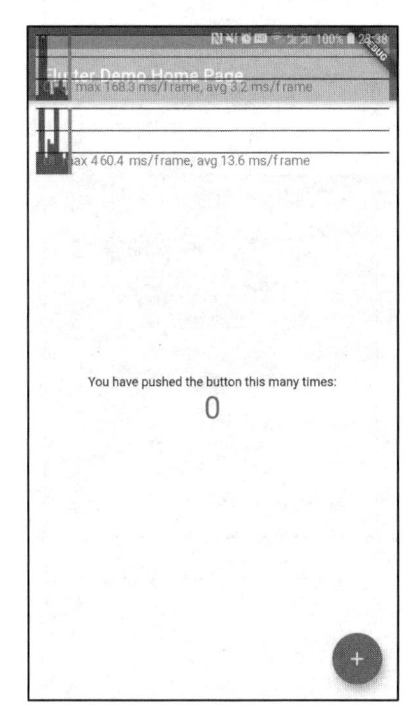

图 16-14　应用程序性能图

在进行布局时，可以使用网格线功能辅助进行对齐验证，使布局更加精准。用法很简单，开启 debugShowMaterialGrid 即可，示例代码如下，效果如图 16-15 所示。

```
class MyApp extends StatelessWidget {
  @override
  Widget build(BuildContext context) {
    return MaterialApp(
```

```
      title: 'Flutter Demo',
      theme: ThemeData(
        primarySwatch: Colors.blue,
      ),
      debugShowMaterialGrid: true,
      home: MyHomePage(title: 'Flutter Demo Home Page'),
    );
  }
}
```

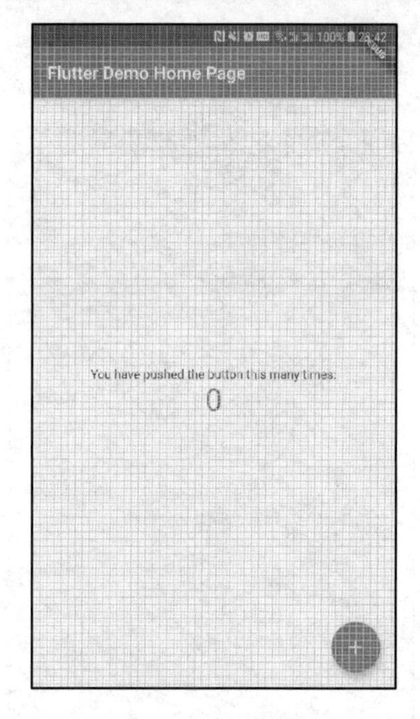

图 16-15　网格线效果

16.1.2　单元测试

Flutter 的单元测试和其他编程语言的单元测试大同小异。项目根目录的 test 目录下默认有一个 widget_test.dart 示例文件，即单元测试文件，内容如下。

```
import 'package:flutter/material.dart';
import 'package:flutter_test/flutter_test.dart';

import 'package:flutter_app/main.dart';

// main()方法为入口函数
```

```
void main() {

  // 单元测试用例方法名
  testWidgets('Counter increments smoke test', (WidgetTester tester) async {
    // 运行应用
    await tester.pumpWidget(MyApp());

    // 证明界面组件显示的是 0，而不是 1
    // 期望找到一个含有 0 这个字符的组件
    expect(find.text('0'), findsOneWidget);
    // 期望没有含有 1 这个字符的组件
    expect(find.text('1'), findsNothing);

    // 模拟触发单击“+”标志
    await tester.tap(find.byIcon(Icons.add));
    await tester.pump();

    // 验证字符已经改变
    // 期望没有含有 0 这个字符的组件
    expect(find.text('0'), findsNothing);
    // 期望找到一个含有 1 这个字符的组件
    expect(find.text('1'), findsOneWidget);
  });
}
```

以上是调用 Flutter 单元测试 API 进行数据和行为验证的测试代码。如果我们用的开发工具是 VSCode，则会看到 Run|Debug 提示按钮，单击这个按钮即可运行单元测试用例，如图 16-16 所示。

```
void main() {
Run | Debug
  testWidgets('Counter increments smoke test', (WidgetTester tester) async {
    // Build our app and trigger a frame.
    await tester.pumpWidget(MyApp());
```

图 16-16　在 VSCode 下运行单元测试用例

16.1.3　辅助工具的使用

本节我们将介绍 Android Studio 下的一些 Flutter 辅助工具，可以在 View->Tool Windows 下查看辅助工具，具体有 Dart Analysis、Flutter Inspector、Flutter Outline、Flutter Performance 等。

Dart Analysis 用于代码的检查和优化分析，如检查代码中的无效常量、变量声明等。

Flutter Inspector 用于可视化和当前页面布局组件树的浏览，其操作栏如图 16-17 所示。

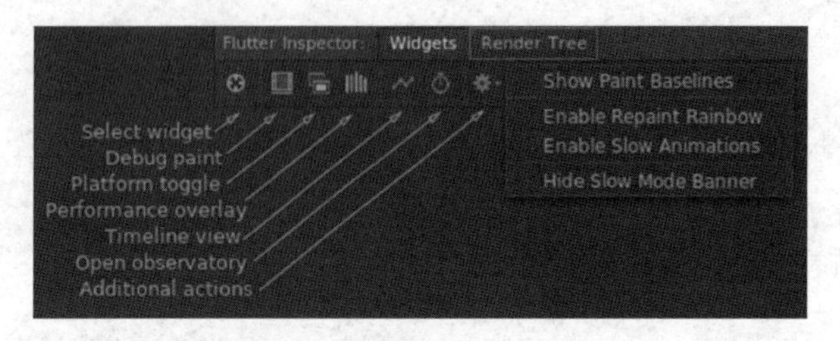

图 16-17　Flutter Inspector 操作栏

单击 Flutter Inspector 操作栏上的“Select widget”选择一个组件，所选组件将在设备和组件树结构中高亮显示。单击真机中的某个组件或 Flutter Inspector 中的某个组件，设备和 Flutter Inspector 界面上都有相应的显示。

Flutter Outline 用于查看代码的结构层级，如图 16-18 所示。

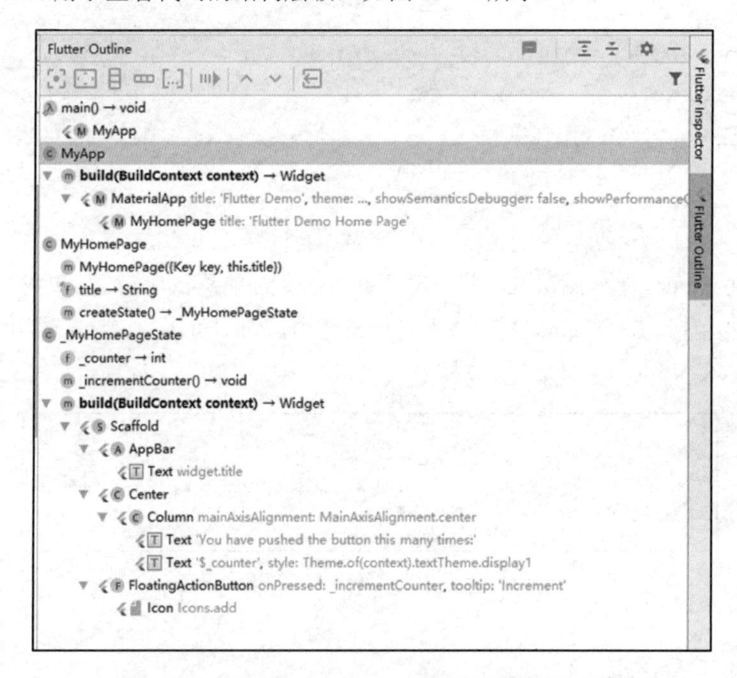

图 16-18　通过 Flutter Outline 查看代码的结构层级

Flutter Performance 主要用于实时性能分析和显示，如显示帧率、显示组件统计信息，如图 16-19 和图 16-20 所示。

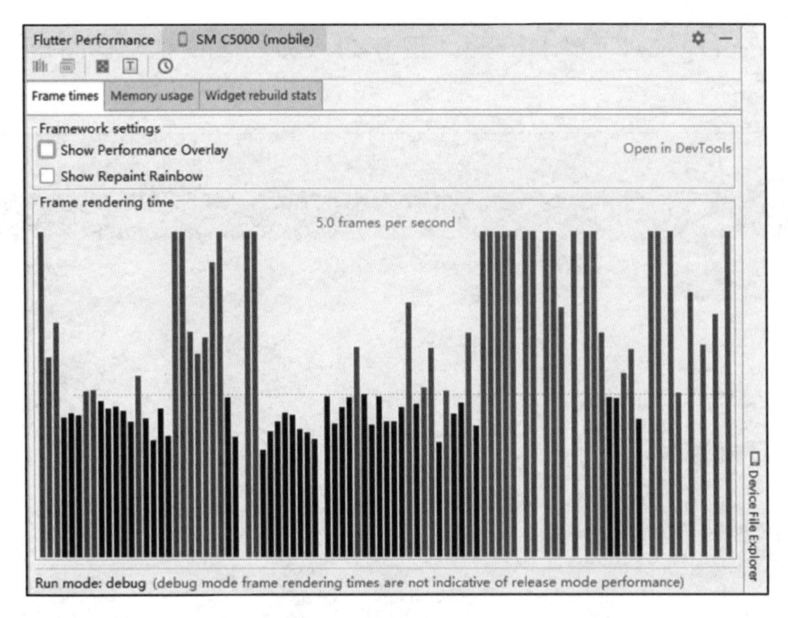

图 16-19　通过 Flutter Performance 显示帧率

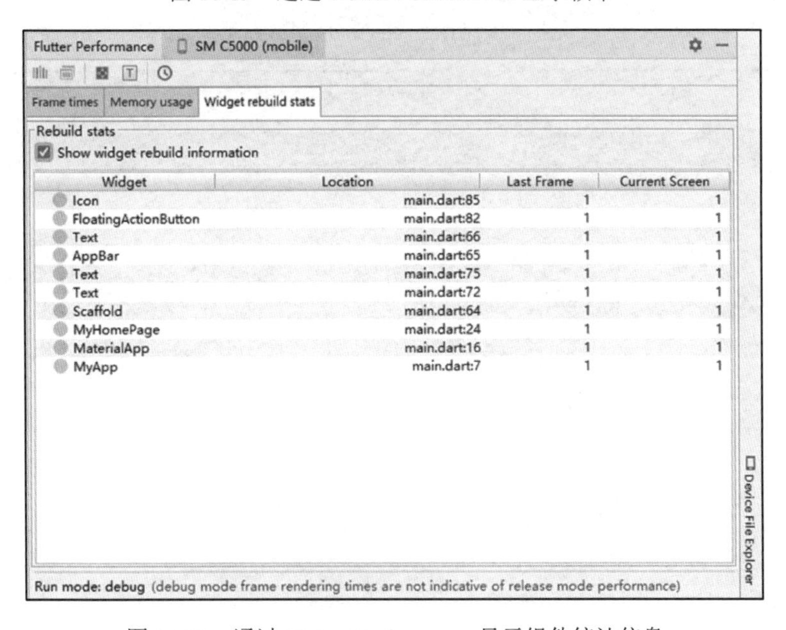

图 16-20　通过 Flutter Performance 显示组件统计信息

当然，Flutter 的辅助工具不止上述几种，还有 Dart DevTools 等。Dart DevTools 集合了很多工具的功能，如 Flutter Inspector、Flutter Timeline、Flutter Memory、Flutter Performance、Flutter

Debugger、Flutter Logging，可谓十分强大。图 16-21 及图 16-22 分别展示了利用 Dart DevTools 调试和查看日志的界面。

图 16-21　Dart DevTools 调试界面

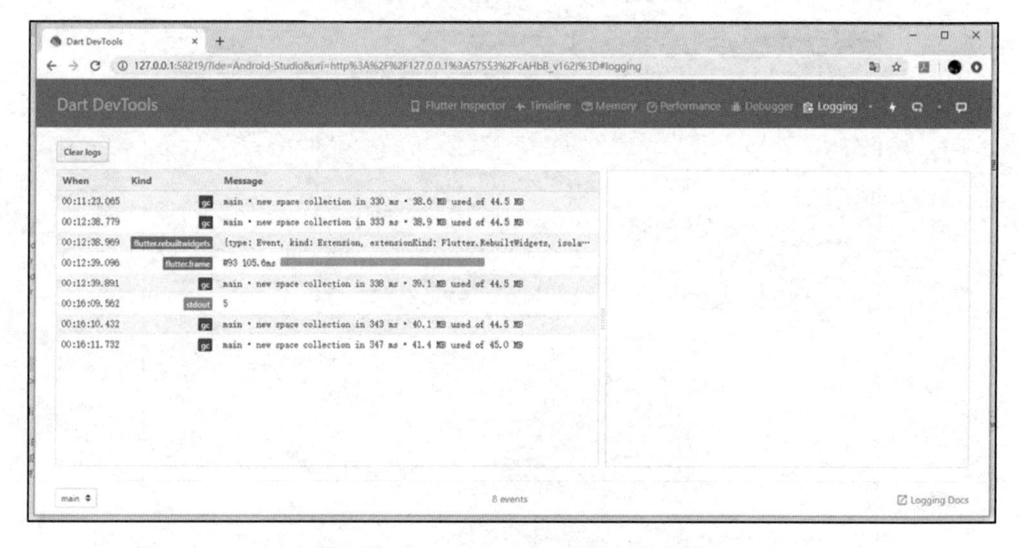

图 16-22　Dart DevTools 查看日志界面

Flutter 中还有许多辅助工具可供使用，能够实现各式各样的功能，这里不再拓展介绍，各位感兴趣的读者可以查阅网络资料，继续学习。

16.2　Flutter Android 应用打包发布

将编写好的应用打包并发布，基本上是 Flutter 应用开发的最后一步。本节我们先来看一下 Android 应用的打包与发布，主要关注点如下。

（1）首先关注应用包名，这是应用的唯一标识，格式类似于 com.google.googlemap，概括来说就是“域名倒着写+应用英文名”。

（2）包名的配置和修改在 build.gradle 配置文件里进行，build.gradle 文件位于项目目录/android/app/下，applicationId 表示包名配置项。

（3）AndroidManifest.xml 主要用于显示是否缺少某些权限。

（4）在项目目录/android/app/src/main/res/下，应将应用图标文件放入对应的 mipmap 文件夹（可以是 mipmap-hdpi、mipmap-mdpi 等，如图 16-23 所示），完成应用图标的替换。

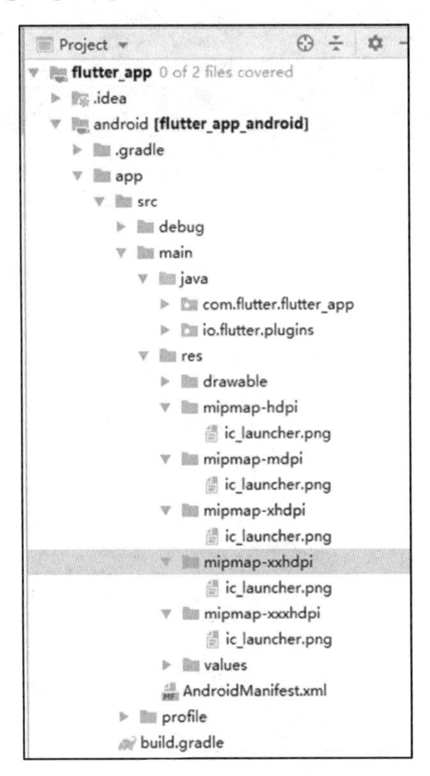

图 16-23　存放应用图标文件的文件夹

（5）最后关注应用签名，应用必须签名后才可以发布和使用。以 Android 为例，制作 Android 应用签名文件时，首先要通过 Android Studio 单独打开 Android module，如图 16-24 所示。

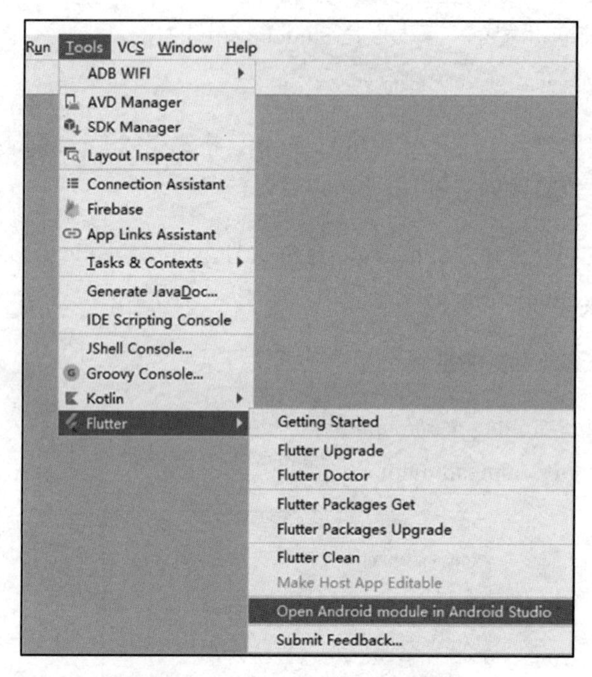

图 16-24　单独打开 Android module

接下来选择 Build-> Generate Signed Bundle/APK，开始使用 Android Studio 制作签名文件，如图 16-25 所示。

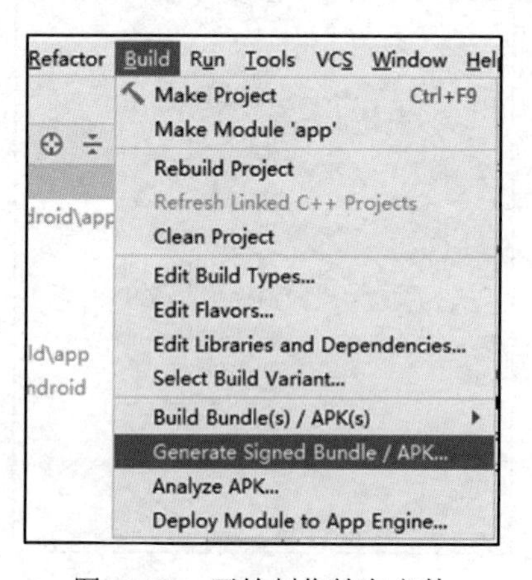

图 16-25　开始制作签名文件

先在弹出的对话框中选择 APK，即选择要进行签名的文件类型，如图 16-26 所示。

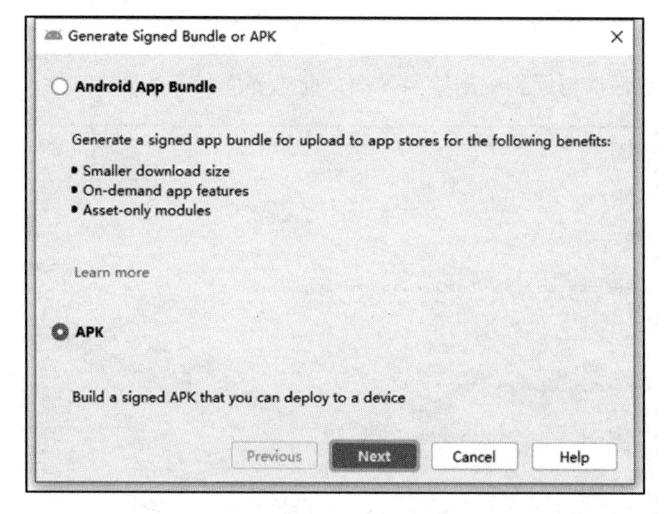

图 16-26　选择要进行签名的文件类型

接下来在 Key store path 中选择签名文件存储路径，输入签名密钥信息。Alias 为签名文件的别名，可以理解为备注。应用签名密钥信息填写界面如图 16-27 所示。

New Key Store

Key store path: D:\flutter_app\flutter_app\flutter.jks

Password: ······ Confirm: ······

Key

Alias: flutterNote

Password: ······ Confirm: ······

Validity (years): 999

Certificate

First and Last Name:

Organizational Unit:

Organization:

City or Locality:

State or Province:

Country Code (XX):

OK Cancel

图 16-27　应用签名密钥信息填写界面

在图 16-27 中，Validity（years）表示有效年限；First and Last Name 表示开发者的名字；

Organizational Unit 表示所在单位；Organization 表示所属组织；City or Locality 表示所在城市；State or Province 表示所在省份；Country Code（XX）表示国家代码。

填写好相关信息后，再回到应用签名界面，此时界面状态如图 16-28 所示。

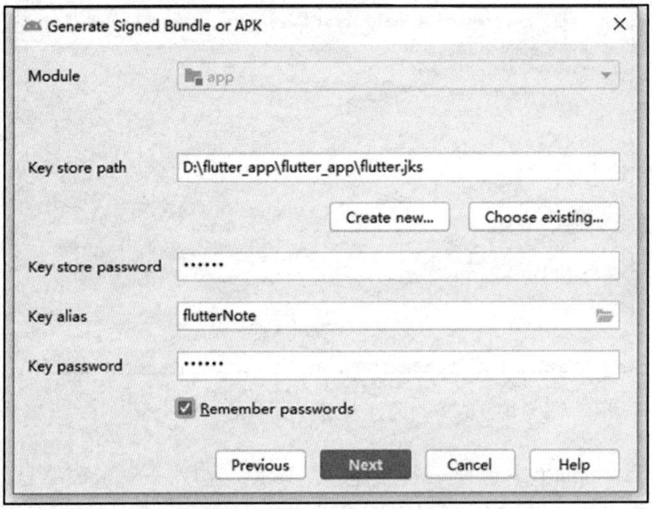

图 16-28 应用签名界面

单击【Next】，选择一个打包模式进行应用 APK 打包，一般选择 release，如图 16-29 所示。

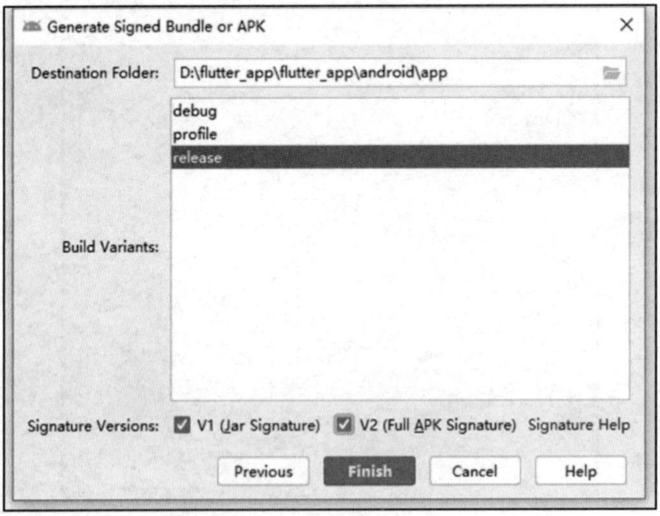

图 16-29 选择打包模式

实际上，进行 Android 应用 APK 打包时无须执行以上步骤，而是要在项目 app 目录中的 build.gradle 里配置签名信息，具体如下。

```
signingConfigs {
        release {
            keyAlias 'flutterNote'
            keyPassword 'key123'
            storeFile file('D:/flutterapp/flutter.jks')
            storePassword 'key123'
            v1SigningEnabled true
            v2SigningEnabled true
        }
    }

    buildTypes {
        debug {
            buildConfigField "boolean", "LOG_DEBUG", "true"
            minifyEnabled false
            proguardFiles getDefaultProguardFile('proguard-android.txt'),
                                                       'proguard-rules.pro'
            signingConfig signingConfigs.debug
        }
        release {
            // 关闭日志输出
            buildConfigField "boolean", "LOG_DEBUG", "false"
            // 关闭代码混淆功能
            minifyEnabled false
            // 优化 Zipalign，减少运行内存消耗
            zipAlignEnabled true
            // 混淆规则文件
            proguardFiles getDefaultProguardFile('proguard-android.txt'),
                                                    'proguard-rules.pro'
            // 签名文件
            signingConfig signingConfigs.release
        }
    }
```

进行以上配置代码后便会根据 debug 版本或 release 版本的签名配置信息实现应用打包时自动签名。接下来在控制台窗口输入打包命令，如下。

```
$ flutter build apk
// 或
$ flutter build apk --release
```

执行以上命令，我们就完成了 Android 应用 APK 的打包。打包好的应用 APK 安装包位于项目目录/build/app/outputs/apk/app-release.apk 下，我们可以将其发布到各个应用商店，实现应用共享。

16.3　Flutter iOS 应用打包发布

本节我们介绍 Flutter iOS 应用的打包发布流程。

首先需要在 macOS 系统下安装好 Android Studio 和 Xcode，其中 Android Studio 可以在 Android 官方网站下载，而 Xcode 在 AppStore 中即可下载。接着在 Android Studio 里安装好 Flutter 和 Dart 插件，下载 Flutter SDK，配置好环境变量。在 macOS 系统下配置环境变量的方法如下。

```
cd ~

// 打开配置文件
open .bash_profile

// 如果 .bash_profile 文件不存在，需要先创建再打开
touch .bash_profile
open .bash_profile

// 配置环境变量
export PATH=${PATH}:/Users/xxx/flutter/bin:$PATH
export PUB_HOSTED_URL=https://pub.flutter-io.cn
export FLUTTER_STORAGE_BASE_URL=https://storage.flutter-io.cn

// 执行配置文件使环境变量生效
source .bash_profile
```

假如 Flutter SDK 目录位于 Users/td 下的 Documents 目录下，那么 Flutter 环境变量的 PATH 应该如下。

```
export PATH=${PATH}:/Users/td/Documents/flutter/bin:$PATH
```

配置好环境变量后，需要检查环境，输入 flutter doctor 命令检查 Flutter 运行环境是否完整，如图 16-30 所示。如不完整，根据提示安装所缺的库或文件即可。

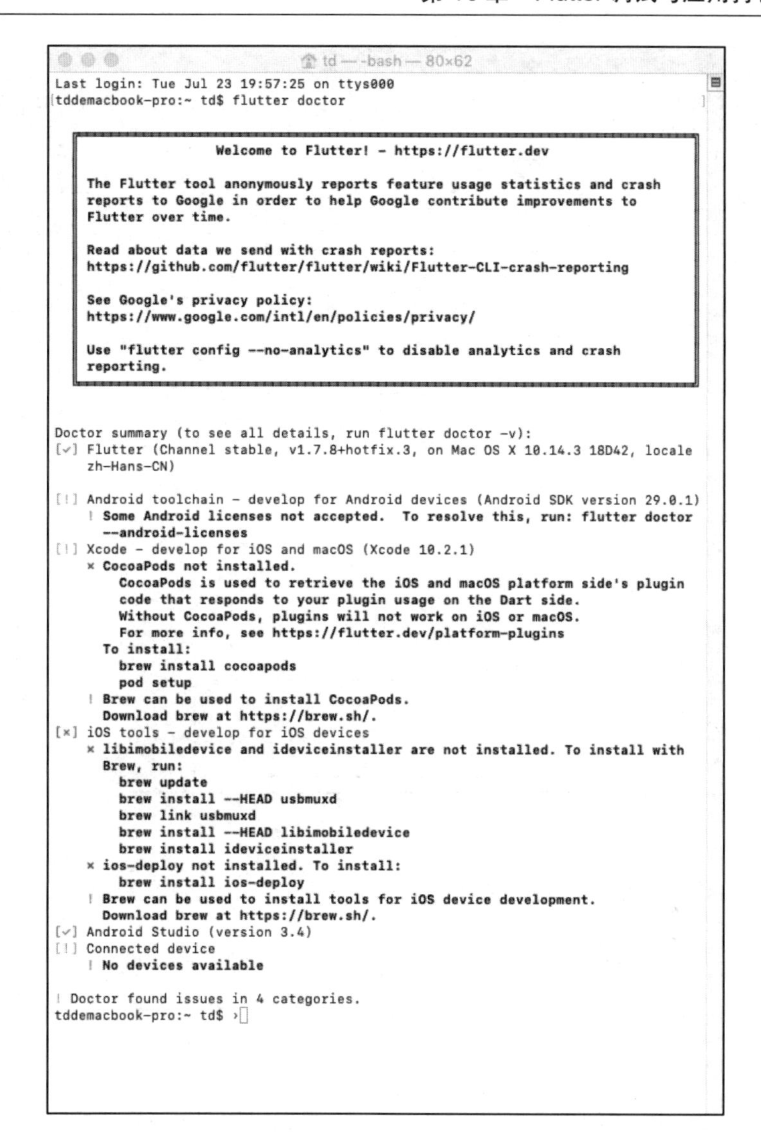

图 16-30　检查环境

完成了 macOS 系统下的环境变量配置并检查通过后，接下来我们进行 iOS 应用打包发布。首先用 Xcode 打开 Flutter 的 iOS 项目目录，打开后可以看到如图 16-31 所示的项目信息。

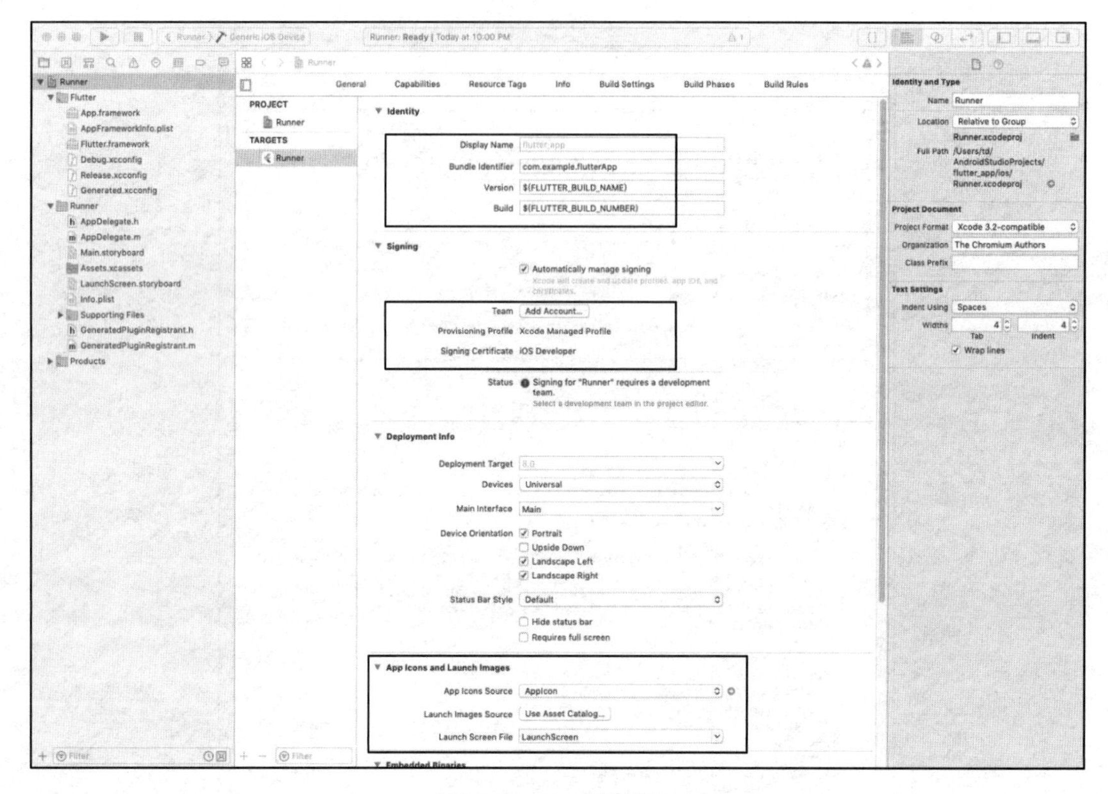

图 16-31　iOS 项目信息

关于图 16-31 中的信息，我们依然只需要关注几个点，具体如下。

- Display Name：要在主屏幕和其他地方显示的应用程序的名称。
- Bundle Identifier：在 iTunes Connect 上注册的 App ID，类似于 Android 应用的唯一包名 com.xxx.xxx。
- Version：应用程序版本名称，类似于 Android 应用的 VersionName。
- Build：应用程序版本号，类似于 Android 应用的 VersionCode。
- Team：选择与 Apple Developer 账户关联的团队，新版 Xcode 要求必须选择 Team。
- App Icons and Launch Images：应用图标设置，当创建新的 Flutter 应用时会同时创建一个 Flutter 占位图标集，此时我们可以用自己的图标替换这些占位图标。替换后，可以运行 flutter run 命令验证应用图标是否已被替换。

回到 iOS 应用的打包与发布步骤。我们需要在 App Store Connect 上注册应用程序。注册应用程序涉及两个步骤：首先注册唯一的 Bundle ID；然后在 App Store Connect 上选择运行设备，创建应用程序。App Store Connect 是创建和管理应用程序的地方，在这里可以填写应用程序名称、添加屏幕截图、设置价格并管理版本、发布到 App Store 和 TestFlight 上。

每一个 iOS 应用程序都与一个 Bundle ID 相关联，注册 Bundle ID 的步骤如下。

1．打开开发者账户的 App IDs 页 。

2．单击“+” 创建 Bundle ID。

3．输入应用程序名称，选择 Explicit App ID，然后输入 Bundle ID。

4．选择应用程序将使用的服务，然后单击“Continue”。

5．确认详细信息，单击“Register”完成 Bundle ID 注册。

在 App Store Connect 上创建应用程序的步骤如下。

1．在浏览器中打开 App Store Connect。

2．在 App Store Connect 登录页面上选择 My App。

3．单击 My App 页面左上角的“+”，选择 New App。

4．填写应用程序的详细信息，在 Platforms 部分确保已选中 iOS（由于 Flutter 目前不支持 tvOS，请不要选中该复选框），然后单击“Create”。

5．跳转到应用程序详细信息页面 App Information，在 General Information 部分填写已注册的 Bundle ID。

以上我们完成了在 App Store Connect 上注册应用程序的工作。如果想正式将应用程序发布到 App Store 或 TestFlight 上，需要准备 release 版本的安装包。首先运行 flutter build ios 来打包 iOS 平台的 release 版本应用，打包完毕后需要将应用归档并上传到 App Store Connect。

如果归档成功，我们会收到一封电子邮件，通知归档已经过验证，可以在 TestFlight 上将应用推送给测试人员。推送应用时，需要将测试人员的邮箱地址添加到 TestFlight 上。

应用测试完毕后，就可以发布到 App Store 了。将应用提交给 App Store 进行审查和发布的步骤如下。

1．从 iTunes 应用程序详情页边栏选择 Pricing and Availability，填写所需信息。

2．从侧边栏选择要发布的应用状态项，是更新应用还是发布第一个版本，对于应用的第一个版本，其状态将为 1.0 Prepare for Submission，选择该版本后要填写必填字段。

3．单击 Submit for Review 完成发布。

经过以上步骤，我们基本完成了 iOS 应用的打包发布，iOS 应用打包发布相对于 Android 应用打包发布麻烦一些，希望各位读者多多实践。

第 17 章

Flutter 拓展：Dart Web

Dart 除了可以用于 Flutter 移动应用开发，还可以用于 Web 开发。本章将拓展介绍 Dart Web 基础知识、Dart Web 开发环境搭建、Dart Web 项目的创建与运行等内容。

17.1 Dart Web 简介

通过 Dart 进行 Web 开发时（注意这里说的是 Dart Web，而不是 Flutter Web），核心思想是用 Dart 替换 JavaScript，也可以说用 Dart 替代 JavaScript 和 JQuery 框架。

我们用 Dart 编写 Web 代码后，借助 dart2js 编译工具会自动将 Dart 文件编译为 JavaScript 文件并运行，只不过在编写时遵循的是 Dart 语法。

Flutter 团队推出了稳定的 Flutter Web SDK 内测版本，基于 Dart 语言实现了 HTML、CSS、JavaScript 组件化开发模式，应用起来十分方便。不过目前还没有发布正式版本，大家可以持续关注。

17.2 Dart Web 环境搭建

关于 Dart Web 环境搭建，Dart 官方提供了英文文档，大家可以在官网查看。另外，Dart 官方也提供了一个在线编程和运行预览的编辑器 DartPad，我们可以使用 DartPad 体验和运行 Dart 程序。DartPad 界面如图 17-1 所示。

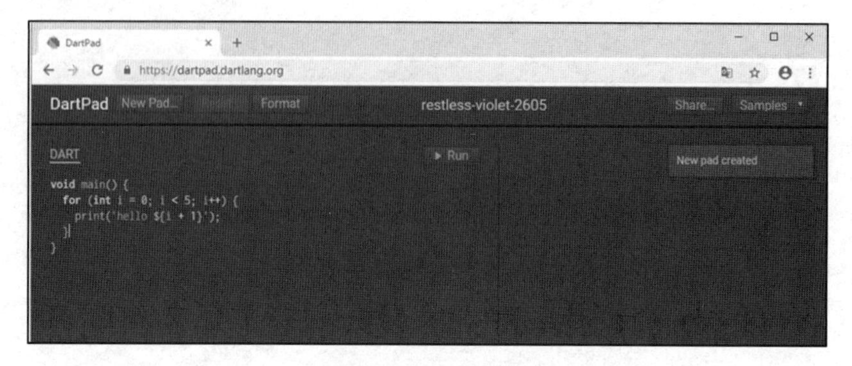

图 17-1　DartPad 界面

17.2.1　下载 Dart SDK

本书是在 Windows 系统下搭建 Dart Web 环境的，首先需要安装 chocolatey。新建一个 chocolatey.bat 文件，将以下命令复制进去保存。运行这个文件就会自动安装 chocolatey 包管理器。

```
@"%SystemRoot%\System32\WindowsPowerShell\v1.0\powershell.exe" -NoProfile -InputForma
t None -ExecutionPolicy Bypass -Command "iex ((New-Object System.Net.WebClient).Downl
oadString('https://chocolatey.org/install.ps1'))" && SET "PATH=%PATH%;%ALLUSERSPROFIL
E%\chocolatey\bin"
```

安装好之后，在 Windows 命令窗口执行如下命令，安装 Dart SDK，界面如图 17-2 所示。

```
C:\> choco install dart-sdk
```

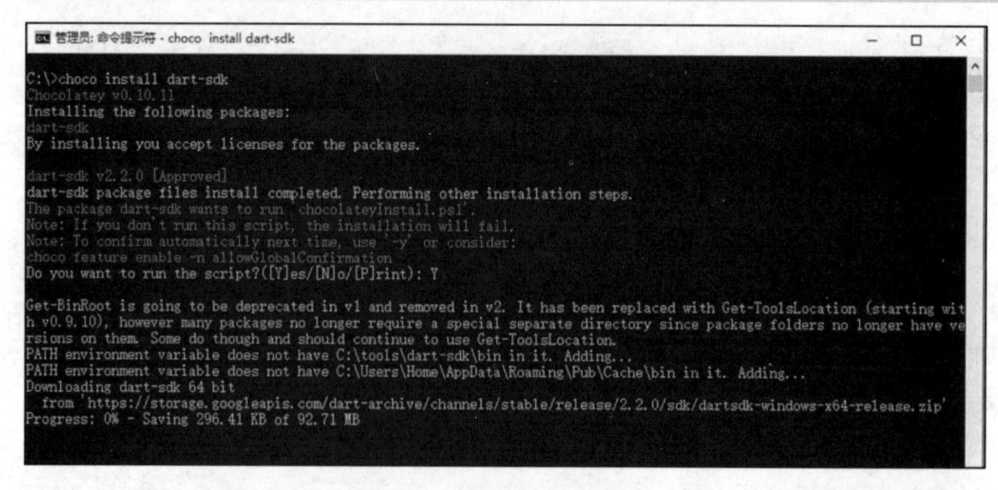

图 17-2　Dart SDK 安装界面

当然，除了通过上述命令安装 Dart SDK，我们也可以安装 Windows 版本的安装包文件，下载地址可以在 Dart 官网获取。图 17-3 显示了 Dart SDK 里包含的部分工具。

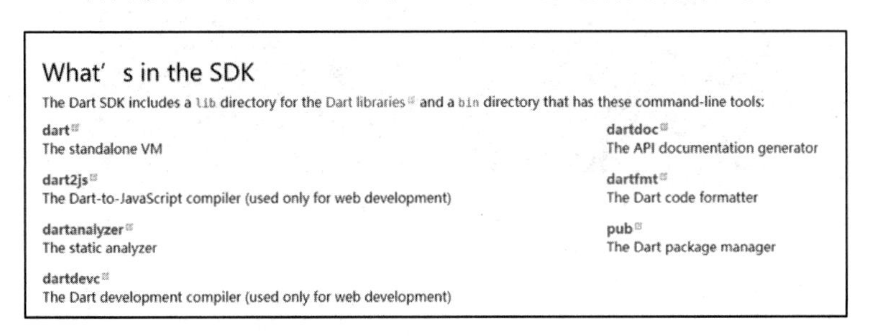

图 17-3　Dart SDK 里包含的部分工具

其中主要的工具是 webdev（图 17-3 中没有标注），主要用来构建和部署 Dart Web 程序；dart2js 工具用于将 Dart 文件编译为 JavaScript 文件；dartdevc 则是一个模块化的将 Dart 文件编译为 JavaScript 文件的工具。

安装完 Dart SDK 后，需要配置 Dart 环境变量，将 Dart SDK 的 bin 目录加入环境变量，如图 17-4 所示。

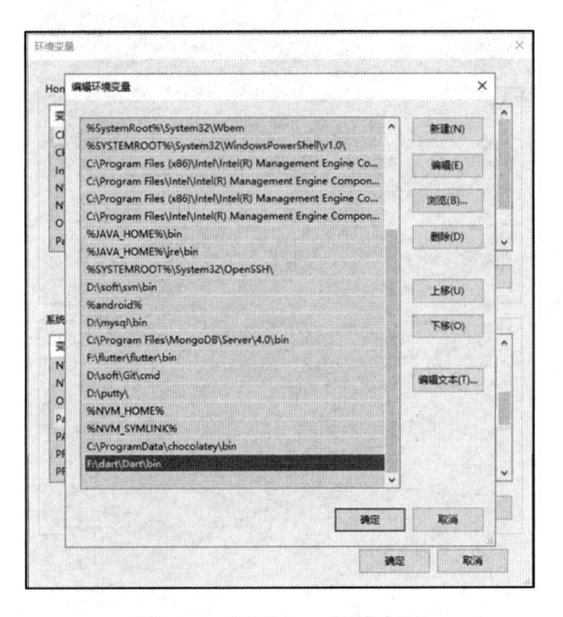

图 17-4　配置 Dart 环境变量

接下来要测试 Dart SDK 环境变量是否已经配置成功。输入如下命令，如果能够正确输出 Dart SDK 的版本号，则表示 Dart 环境变量配置成功，如图 17-5 所示。

```
dart --version
```

```
命令提示符
Microsoft Windows [版本 10.0.17134.648]
(c) 2018 Microsoft Corporation。保留所有权利。

C:\Users\Home>dart --version
Dart VM version: 2.2.0 (Tue Feb 26 15:04:32 2019 +0100) on "windows_x64"

C:\Users\Home>
```

图 17-5　Dart 环境变量配置成功

17.2.2　下载开发工具

官方推荐的开发工具是 WebStorm，当然也可以使用 VSCode 进行开发。在下载开发工具前，需要先通过如下命令安装 webdev 和 stagehand，安装窗口如图 17-6 所示。

```
> pub global activate webdev
> pub global activate stagehand
```

```
命令提示符
Microsoft Windows [版本 10.0.17134.648]
(c) 2018 Microsoft Corporation。保留所有权利。

C:\Users\Home>pub global activate webdev
Package webdev is currently active at version 1.0.1.
Resolving dependencies...
+ args 1.5.1
+ charcode 1.1.2
+ collection 1.14.11
+ io 0.3.3
+ meta 1.1.7
+ path 1.6.2
+ pub_semver 1.4.2
+ source_span 1.5.5
+ stack_trace 1.9.3
+ string_scanner 1.0.4
+ term_glyph 1.1.0
+ webdev 1.0.1
+ yaml 2.1.15
Precompiling executables...
Precompiled webdev:bin\webdev.
Installed executable webdev.
Activated webdev 1.0.1.

C:\Users\Home>pub global activate stagehand
Package stagehand is currently active at version 3.2.0.
Resolving dependencies...
+ args 1.5.1
+ async 2.0.8
+ charcode 1.1.2
+ collection 1.14.11
+ http 0.12.0+1
+ http_parser 3.1.3
+ meta 1.1.7
+ path 1.6.2
+ source_span 1.5.5
+ stagehand 3.2.0
+ string_scanner 1.0.4
+ term_glyph 1.1.0
+ typed_data 1.1.6
+ usage 3.4.1
Precompiling executables...
Precompiled stagehand:bin\stagehand.
Installed executable stagehand.
Activated stagehand 3.2.0.
```

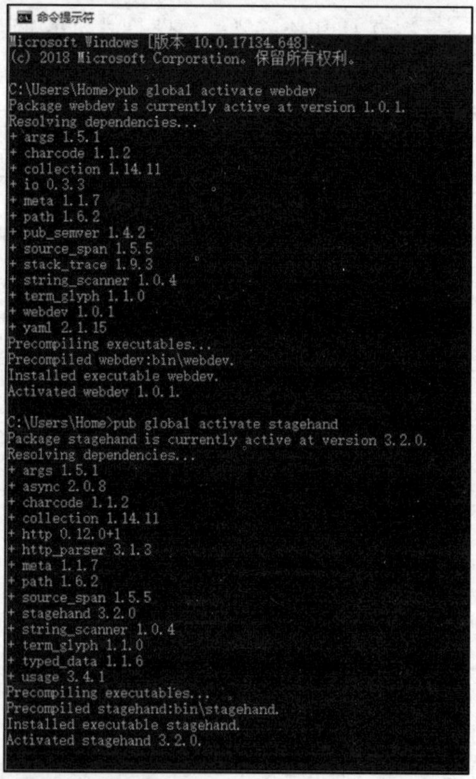

图 17-6　安装 webdev 和 stagehand 的窗口

注意：如果想运行 Dart 2 以下的版本，WebStorm 的版本需要是 2018.1.3 及以上。当然，现在基本都用 Dart 2 及以上版本进行开发。

接下来安装 WebStorm。我们需要在 jetbrains 官方网站下载 WebStorm IDE，按照提示步骤进行安装，界面如图 17-7 所示。开发环境和开发工具均配置成功，接下来就可以创建一个 Dart Web 项目了。

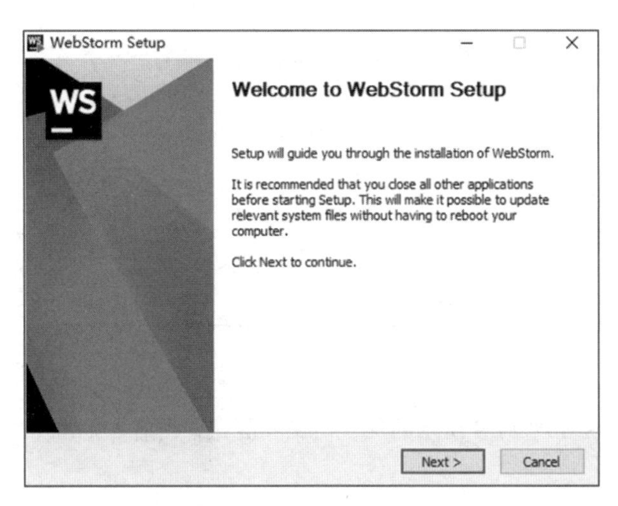

图 17-7　WebStorm 安装界面

17.3　创建一个 Dart Web 项目

创建 Dart Web 项目的方式有两种：通过命令行创建、通过 WebStorm IDE 创建。我们先来看一下通过命令行创建 Dart Web 项目的方法，如下。

```
> mkdir quickstart
> cd quickstart
> stagehand web-simple
> pub get
```

通过 WebStorm IDE 创建 Dart Web 项目相对复杂，首先要创建新项目，如图 17-8 所示。

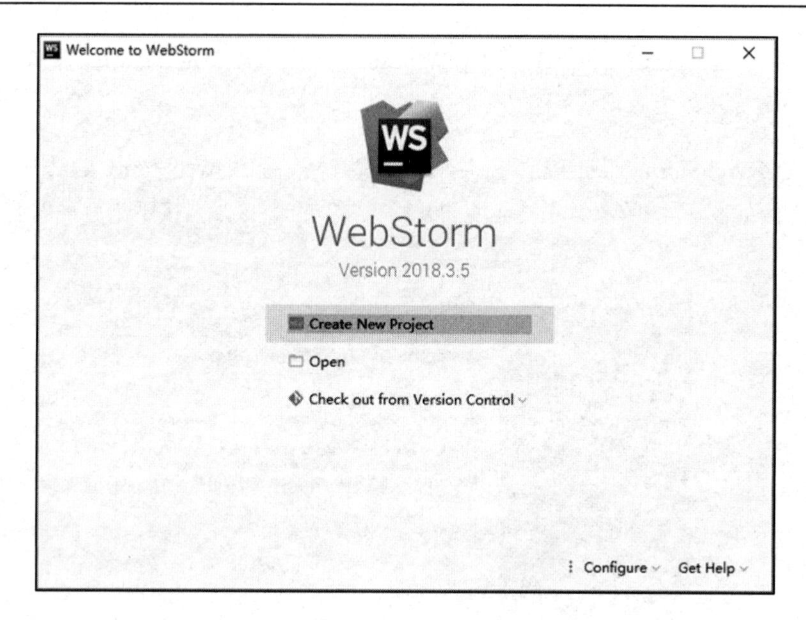

图 17-8　创建新项目

选择 Dart 项目，在如图 17-9 所示的项目信息界面完善内容，单击【CREATE】按钮，完成项目创建。

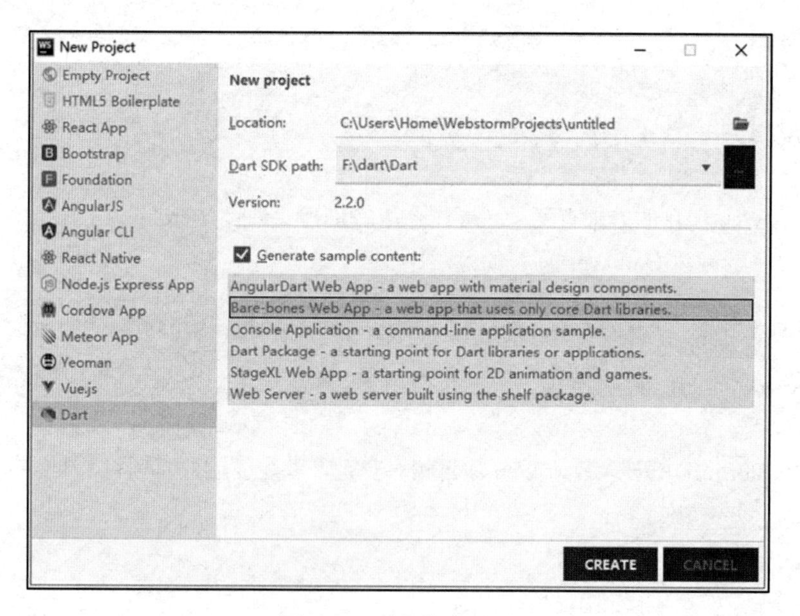

图 17-9　项目信息界面

Dart Web 项目结构如图 17-10 所示。

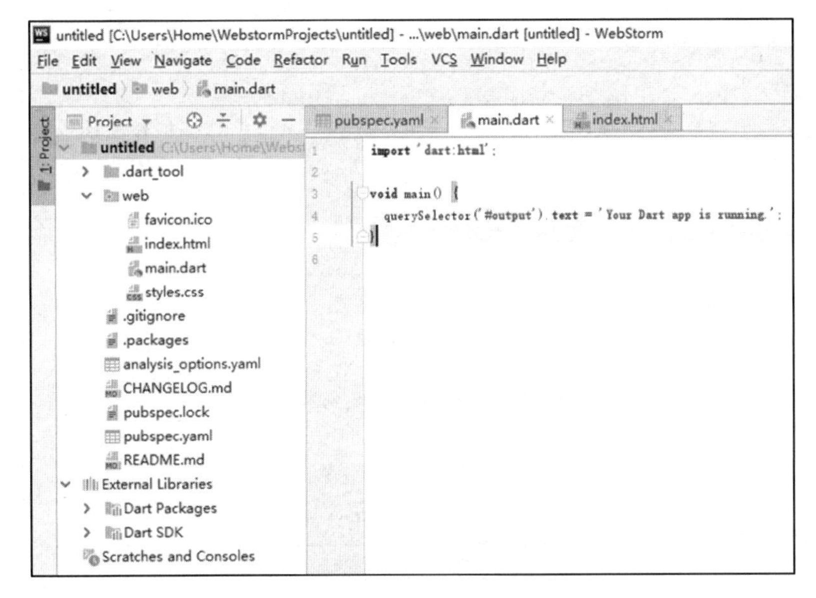

图 17-10　Dart Web 项目结构

创建完毕后，我们就可以运行项目了，如图 17-11 所示。

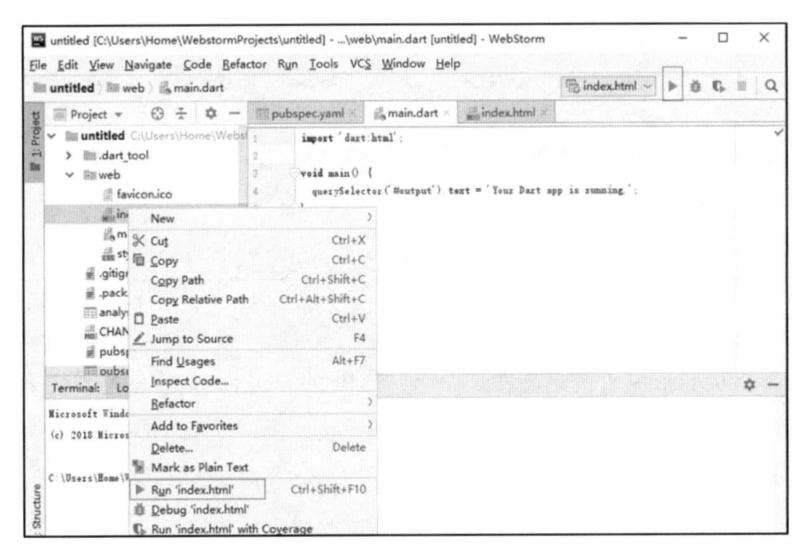

图 17-11　运行项目

运行后，如果可以看到控制台显示的日志（见图 17-12），即不出现报错现象，就可以访问页面了。

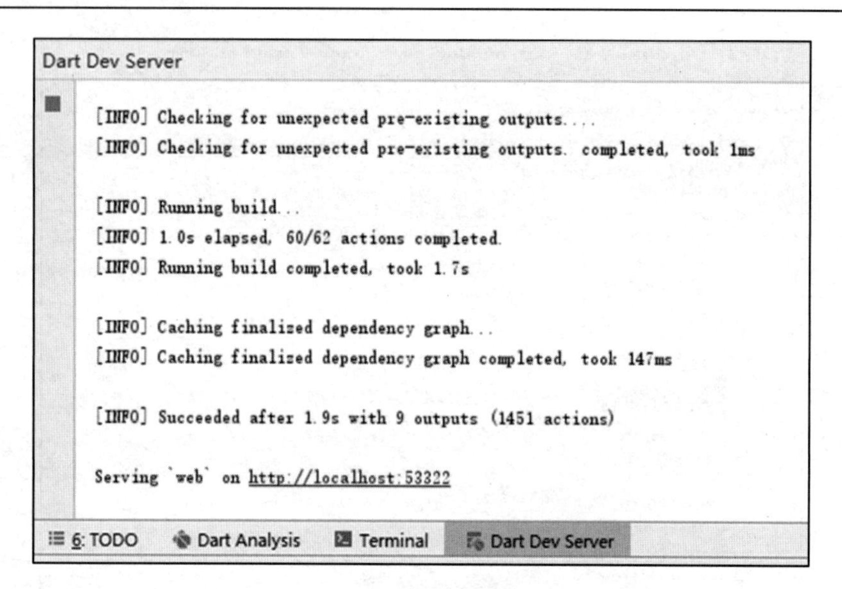

图 17-12　控制台显示的日志

用浏览器打开相应网址预览页面，页面效果如图 17-13 所示。

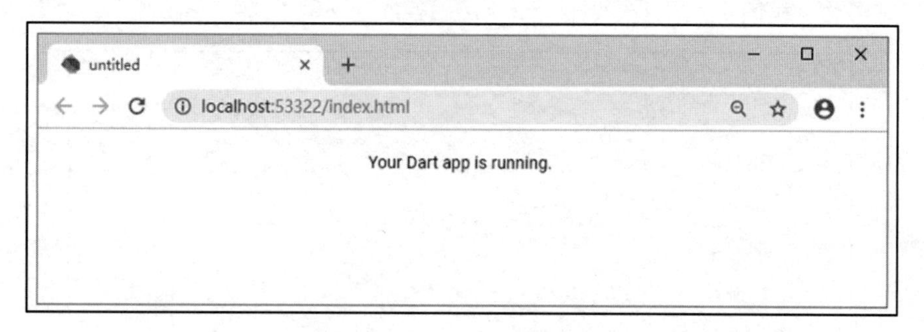

图 17-13　页面效果

可以看出，这里 Dart 文件的功能就是操作 HTML 的 DOM 树，也就是替代 JavaScript 的原始用法。以下是操作 DOM 的 Dart 语法，简单清晰，IDE 也有代码自动补全功能。

```
import 'dart:html';

void main() {
  querySelector('#output').text = 'Your Dart app is running.';
}
```

而 HTML 部分内容和标准的 HTML 规范是一模一样的，这里没有改动，代码如下。

```
<!DOCTYPE html>
```

```
<html>
<head>
    <meta charset="utf-8">
    <meta http-equiv="X-UA-Compatible" content="IE=edge">
    <meta name="viewport" content="width=device-width, initial-scale=1.0">
    <meta name="scaffolded-by" content="https://github.com/google/stagehand">
    <title>untitled</title>
    <link rel="stylesheet" href="styles.css">
    <link rel="icon" href="favicon.ico">
    <script defer src="main.dart.js"></script>
</head>

<body>

  <div id="output"></div>

</body>
</html>
```

关于 Dart Web 环境搭建及项目创建就介绍这么多，大家可以动手实践，辅助理解。

17.4　编写第一个 Dart Server

学习完 Dart Web 相关知识后，我们再来拓展学习 Dart Server 相关知识。Dart Server 的环境配置和工具安装同 Dart Web 一样，这里不再赘述。最简单的 Dart Server 项目包括如下几个部分。

- 一个以.dart 为后缀的源文件。
- 一个顶层的 main()方法入口函数。
- 名为 main.dart 的入口文件，其中入口函数名固定为 main()。

Dart Server 项目结构如图 17-14 所示，其中几个比较常见的目录功能如下。

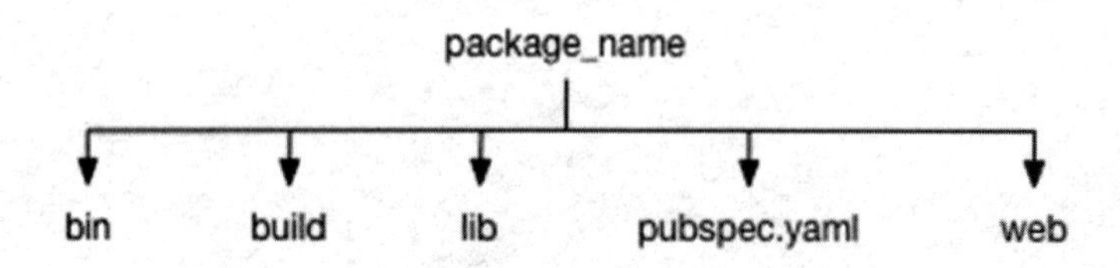

图 17-14 Dart Server 项目结构

- bin 目录：存放命令行应用程序文件，其中一个文件中必须有 main()入口函数。
- lib 目录：存放应用的附属代码或库文件。
- pubspec.yaml：存放应用的配置和描述信息文件，和 Flutter 的 pubspec.yaml 一样。

我们先来看一个比较简单的 Dart Server 项目（Command-line apps）的创建。Dart Server 项目是通过命令行执行的程序，通常用于为 Web 应用程序提供服务器端支持。这里我们使用 WebStorm IDE 创建 Dart Server 项目，如图 17-15 所示。

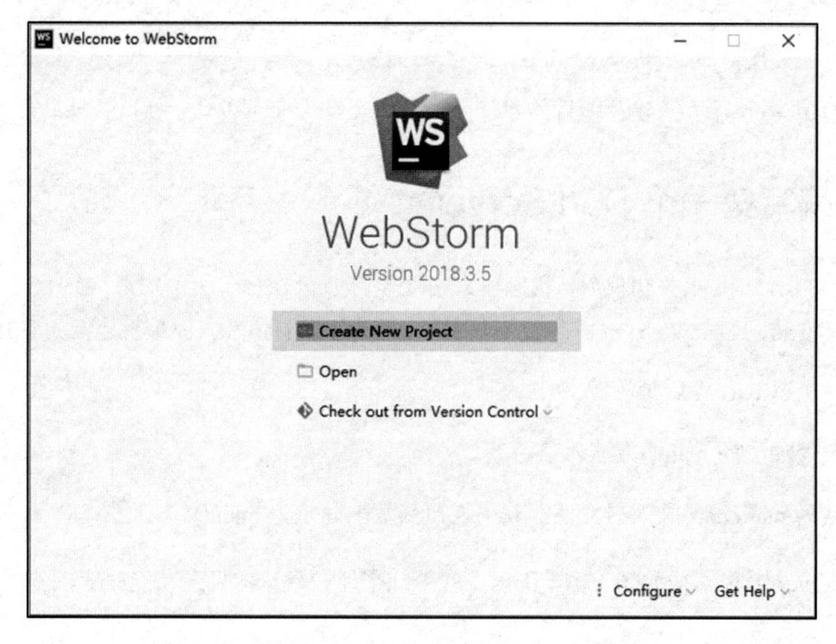

图 17-15 使用 WebStorm IDE 创建 Dart Server 项目

选择 Dart 项目，再选择 Web Server，完善项目信息，如图 17-16 所示。单击【CREATE】即可完成创建。

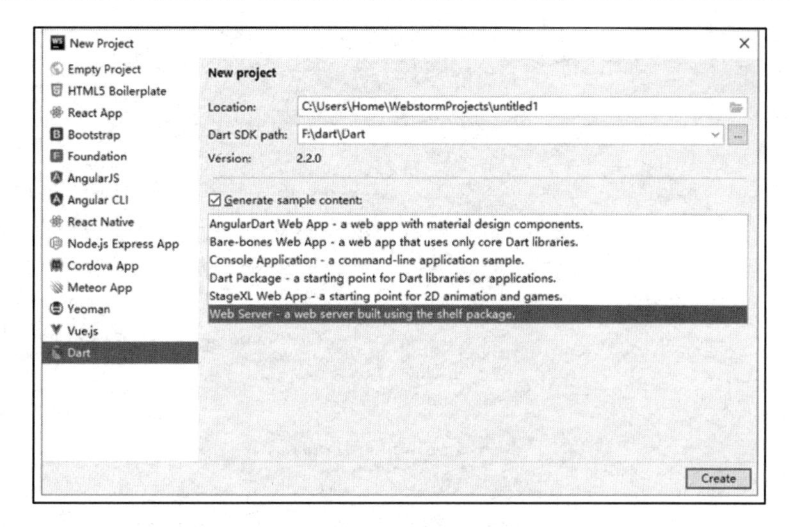

图 17-16　完善项目信息

也可以选择 Bare-bones Web App 这一项，如图 17-17 所示。选择这一项会自动在项目目录中创建一个 bin 目录。

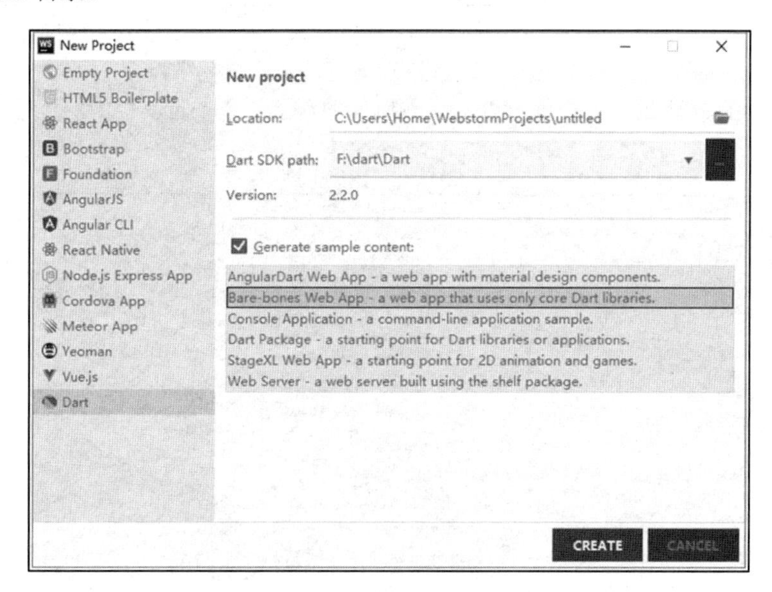

图 17-17　选择 Bare-bones Web App

创建后的 Dart Server 项目结构目录如图 17-18 所示。

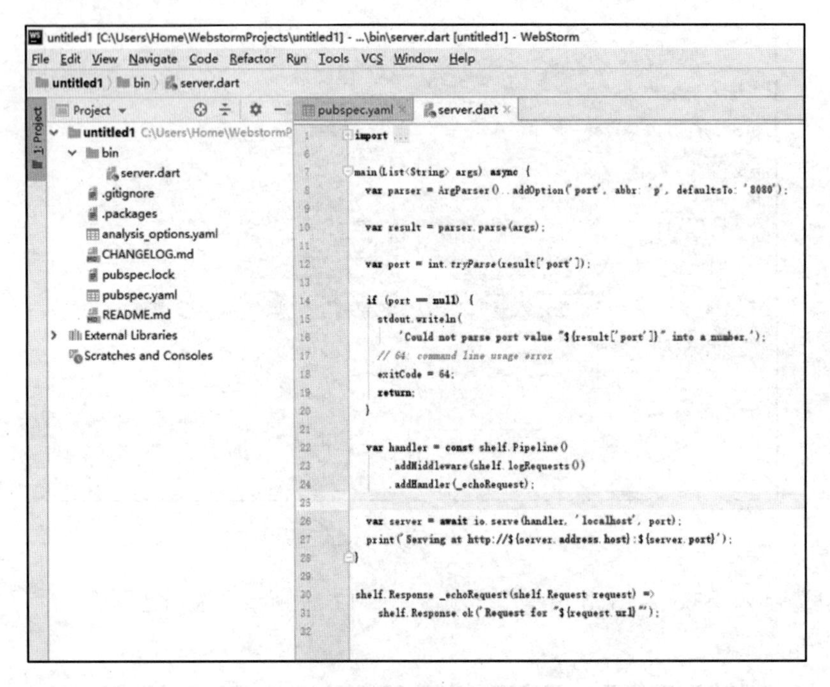

图 17-18　Dart Server 项目结构目录

完成项目代码逻辑编写后，就可以编译并运行项目了，单击工具栏中的运行按钮即可，如图 17-19 所示。

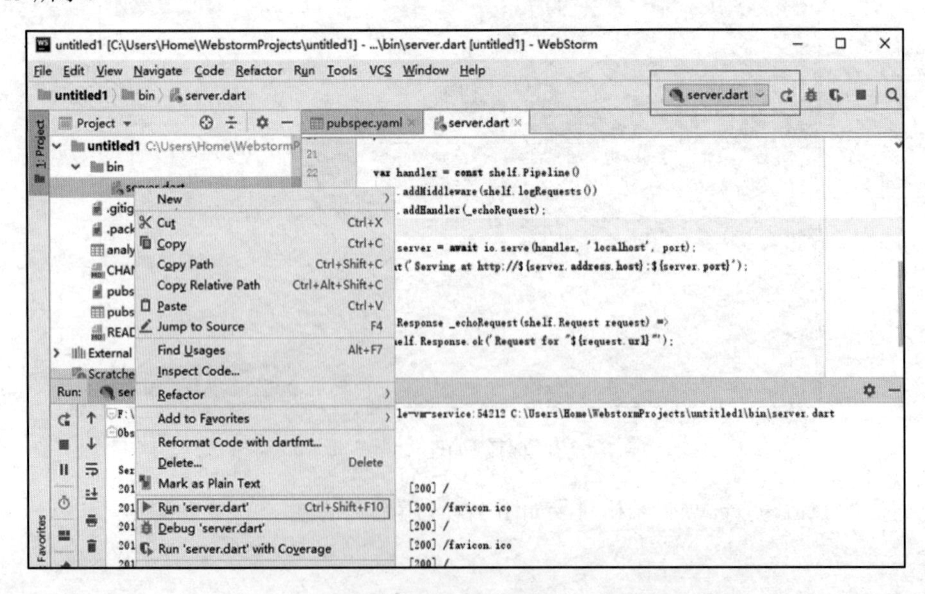

图 17-19　运行项目

运行后，可以看到控制台显示日志，如果看到 Dart Server 启动成功的日志（如图 17-20 所示）就表示可以访问页面了，页面效果如图 17-21 所示。

```
F:\dart\Dart\bin\dart.exe --enable-asserts --enable-vm-service:54212 C:\Users\Home\WebstormProjects\untitled1\bin\server.dart
Observatory listening on http://127.0.0.1:54212/

Serving at http://localhost:8080
2019-03-14T22:37:14.794893 0:00:00.008343 GET    [200] /
2019-03-14T22:37:14.903602 0:00:00.000171 GET    [200] /favicon.ico
2019-03-14T22:37:17.554877 0:00:00.000194 GET    [200] /
2019-03-14T22:37:17.567872 0:00:00.000172 GET    [200] /favicon.ico
2019-03-14T22:37:18.261098 0:00:00.000157 GET    [200] /
2019-03-14T22:37:18.274063 0:00:00.000212 GET    [200] /favicon.ico
2019-03-14T22:37:48.674824 0:00:00.000127 GET    [200] /dafdas
2019-03-14T22:37:48.687789 0:00:00.000232 GET    [200] /favicon.ico
2019-03-14T22:37:53.506690 0:00:00.000142 GET    [200] /
2019-03-14T22:37:53.518658 0:00:00.000145 GET    [200] /favicon.ico
```

图 17-20　Dart Server 启动成功的日志

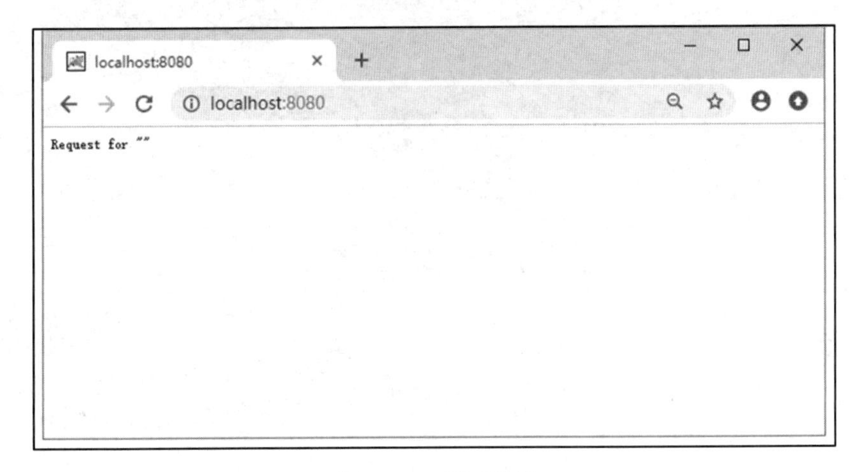

图 17-21　页面效果

这样就完成了一个 Dart Server 项目的创建与运行，接下来我们可以编写一些 Server 脚本程序工具或后台服务，实现其他后端编程语言的功能，例如 Node.js、Java 等的后端服务功能。

官方给出的详细 HTTP Server 后端应用（HTTP Clients & Servers）英文文档可以在 Dart 官方文档中查看。官方 HTTP Server 后端应用默认示例代码如下，大家可以参考。

```
import 'dart:io';

import 'package:args/args.dart';
```

```
import 'package:shelf/shelf.dart' as shelf;
import 'package:shelf/shelf_io.dart' as io;

main(List<String> args) async {
  var parser = ArgParser()..addOption('port', abbr: 'p', defaultsTo: '8080');

  var result = parser.parse(args);

  var port = int.tryParse(result['port']);

  if (port == null) {
    stdout.writeln(
        'Could not parse port value "${result['port']}" into a number.');
    // 64: command line usage error
    exitCode = 64;
    return;
  }

  var handler = const shelf.Pipeline()
      .addMiddleware(shelf.logRequests())
      .addHandler(_echoRequest);

  var server = await io.serve(handler, 'localhost', port);
  print('Serving at http://${server.address.host}:${server.port}');
}

shelf.Response _echoRequest(shelf.Request request) =>
    shelf.Response.ok('Request for "${request.url}"');
```

第 18 章 Flutter 实战

在前面的章节中，我们将关于 Flutter 开发的内容进行了全面介绍，本章我们将基于前面学习的知识进行实战：编写一个简易备忘录应用，以及编写一个 TV 应用。通过这两个示例，希望大家巩固之前学过的知识，高效查缺补漏。

18.1 编写一个备忘录应用

备忘录应用是一个典型示例，里面涉及列表布局的编写、数据库增删改查操作、路由操作、通信实现、复杂布局的编写等，基本能涵盖所有常用的 Flutter 知识点。

18.1.1 知识整理

在进行综合实战前，先来整理编写备忘录应用可能涉及的知识点，具体如下。

- 引导页（PageView）
- 顶部 ToolBar（AppBar）
- 列表（CustomScrollView）
- 日历（第三方库 flutter_custom_calendar）
- 权重（Flexible、Expanded）
- 导航（CupertinoTabBar）
- 弹窗（BottomSheet、SnackBar）

- 输入框（TextField）
- 通知刷新（EventBut）
- 路由
- 下拉刷新（RefreshIndicator）
- 数据库（官方数据库 sqflite）
- 时间格式化
- 生命周期监听
- 返回键拦截
- List 集合操作
- 复杂布局

18.1.2　应用编写

我们要实现的备忘录应用的功能是，打开备忘录后显示引导页，应用首页底部显示导航栏，提供三个页面进行切换，具备备忘录列表实时更新功能、增删改查功能、搜索功能、复制粘贴功能，应用中的数据是存储在本地数据库内的。

根据以上功能，我们可以总结出以下关键点。

- 引导页：通过 PageView 可以实现一个可滑动的引导页。
- 应用页面框架：包括一个底部导航栏，含三个可切换页面，切换时保持页面数据状态，页面可使用 Scaffold 脚手架快速搭建。
- 增删改查功能：数据库操作，数据存储在手机本地数据库中。
- 搜索功能：主要用于列表展示。
- 列表实时更新功能：当新增或修改备忘录后，返回列表首页会实时更新。
- 复制粘贴功能：实现列表文字的复制与粘贴，涉及 Flutter 的复制粘贴功能。

该应用中使用到的第三方库如下。

- sqflite：负责实现增删改查功能，对数据库进行操作。
- path_provider：负责读取目录路径。
- oktoast：实现应用 Toast 信息提示功能。
- event_bus：实现列表实时更新功能，即通知刷新功能。
- flutter_custom_calendar：实现日历功能。

应用的 pubspec.yaml 配置文件的依赖库配置信息大致如下。

```
dependencies:
  flutter:
    sdk: flutter
  sqflite: ^1.1.6+3
  cupertino_icons: ^0.1.2
  path_provider: ^1.1.0
  oktoast: ^2.2.0
  event_bus: ^1.1.0
  flutter_custom_calendar:
    git:
      url: https://github.com/LXD312569496/flutter_custom_calendar.git
```

由于这个备忘录应用的功能比较完善，所以这里挑选其中比较重要的技术点进行代码讲解。其他代码可以在 GitHub 上搜索 flutter_notes，下载并查看。先来看一下应用的目录结构，如图 18-1 所示。

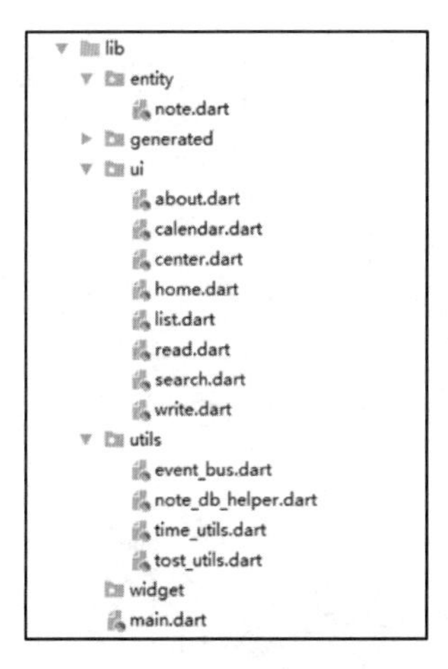

图 18-1　应用的目录结构

接下来我们实现引导页，效果如图 18-2 所示。

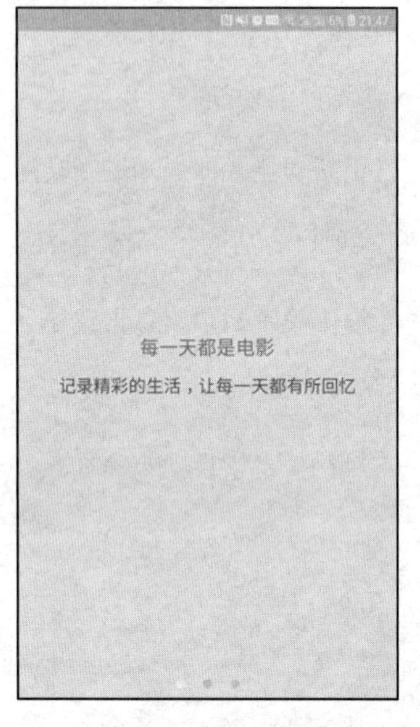

图 18-2　引导页效果

引导页主要由三个可左右滑动的页面组成，并且有游标指示，滑动到最后一页时可以通过特定手势进入应用首页。具体的实现代码如下（部分非核心代码省略）。

```
// 用 PageView 实现引导页
… …
class MyHomePage extends StatefulWidget {
  @override
  _MyHomePageState createState() => _MyHomePageState();
}

class _MyHomePageState extends State<MyHomePage> {
  // 默认选中第一个页面
  int _selectedIndex = 0;
  var _pageController = new PageController(initialPage: 0);

  @override
  void initState() {
    super.initState();
    _pageController.addListener(() {
```

```
    print(_pageController.position);
  });
}

@override
Widget build(BuildContext context) {
  return Scaffold(
    body: _buildBody(),
  );
}
// 构建页面主体
_buildBody() {
  // 用 SafeArea 包裹，保证不会在刘海屏等屏幕下被剪裁
  return SafeArea(
      child: Stack(
    children: <Widget>[
      // 用 PageView 实现引导页页面切换
      PageView(
        // 监听控制类
        controller: _pageController,
        onPageChanged: _onSelectChanged,
        children: <Widget>[
          // 引导页第一个页面
          Container(
            color: Color.fromRGBO(232, 229, 222, 1),
            child: Column(
              crossAxisAlignment: CrossAxisAlignment.center,
              mainAxisAlignment: MainAxisAlignment.center,
              children: <Widget>[
                Text(
                  '每一天都是电影',
                  style: TextStyle(
                      color: Color.fromRGBO(132, 112, 101, 1), fontSize: 22),
                ),
                SizedBox(
                  height: 10,
                ),
                Text(
                  '记录精彩的生活，让每一天都有所回忆',
```

```
                style: TextStyle(
                    color: Color.fromRGBO(66, 84, 94, 1), fontSize: 20),
              )
            ],
          ),
        ),
        // 引导页第二个页面，这里省略
        ……
        // 引导页第三个页面，这里省略
        … …
      ],
    ),
    // 绘制引导页的三个圆点指示器
    Align(
      alignment: FractionalOffset.bottomCenter,
      child: Row(
        mainAxisAlignment: MainAxisAlignment.center,
        children: <Widget>[
          Container(
              width: 10,
              height: 10,
              margin: EdgeInsets.all(10),
              decoration: BoxDecoration(
                  shape: BoxShape.circle,
                  color: (_selectedIndex == 0)
                      ? Colors.white70
                      : Colors.black12)),
           // 第二个和第三个省略
        ],
      ),
    )
  ],
));
}

void _onSelectChanged(int index) {
  setState(() {
    _selectedIndex = index;
  });
```

```
  }

  // 切换选中的页面
  void setPageViewItemSelect(int indexSelect) {
    _pageController.animateToPage(indexSelect,
        duration: const Duration(milliseconds: 300), curve: Curves.ease);
  }

  toPage() {
    // 跳转并关闭当前页面
    Navigator.pushAndRemoveUntil(
      context,
      MaterialPageRoute(builder: (context) {
        return HomePage();
      }),
      (route) => route == null,
    );
  }
}
```

以上为打开备忘录应用时的引导页实现代码，接下来我们看一下应用首页如何实现。应用首页效果如图 18-3 所示。

图 18-3　应用首页效果

应用首页由顶部的 AppBar、底部的菜单栏和中间的列表组成，结构清晰合理。这里我们会借助 Scaffold 脚手架来快速搭建这个页面。首页的基本框架和结构代码如下。

```
// 只给出核心代码，其余部分省略
… …
class HomePageState extends State<HomePage> {
  // 默认选中第一个页面
  int _selectedIndex = 0;
  var _pageController = new PageController(initialPage: 0);
  int last = 0;
  int index = 0;

  NoteDbHelper noteDbHelpter;

  @override
  void initState() {
    super.initState();
    noteDbHelpter = NoteDbHelper();
    getDatabasesPath().then((string) {
      String path = join(string, 'notesDb.db');
      noteDbHelpter.open(path);
    });
    _pageController.addListener(() {});
  }

  // 返回键拦截执行方法
  Future<bool> _onWillPop() {
    int now = DateTime.now().millisecondsSinceEpoch;
    print(now - last);
    if (now - last > 1000) {
      last = now;
      Toast.show("再按一次返回键退出");
      return Future.value(false); // 不退出
    } else {
      return Future.value(true); // 退出
    }
  }

  @override
```

```
Widget build(BuildContext context) {
  // 用 WillPopScope 包裹
  return WillPopScope(
      // 编写 onWillPop 逻辑
      onWillPop: _onWillPop,
      child: Material(
        child: SafeArea(
            child: Scaffold(
          appBar: PreferredSize(
              // 通过 Offstage 来控制 AppBar 的显示与隐藏
              child: Offstage(
                offstage: _selectedIndex == 2 ? true : false,
                child: AppBar(
                  backgroundColor: Color.fromRGBO(244, 244, 244, 1),
                  title: Text('备忘录'),
                  primary: true,
                  automaticallyImplyLeading: false,
                  actions: <Widget>[
                    IconButton(
                      icon: Icon(Icons.search),
                      tooltip: '搜索',
                      onPressed: () {
                        Navigator.push(context,
                            MaterialPageRoute(builder: (context) {
                          return SearchPage(
                            noteDbHelpter: noteDbHelpter,
                          );
                        }));
                      },
                    ),
                    IconButton(
                      icon: Icon(Icons.add),
                      tooltip: '写日记',
                      onPressed: () {
                        Navigator.push(context,
                            MaterialPageRoute(builder: (context) {
                          return WritePage(
                            noteDbHelpter: noteDbHelpter,
                            id: -1,
```

```
                  );
                }));
              },
            ),
          ],
        ),
      ),
      preferredSize:
          Size.fromHeight(MediaQuery.of(context).size.height * 0.07)),
// 绑定数据
body: SafeArea(
    child: PageView(
  // 监听控制类
  controller: _pageController,
  onPageChanged: _onItemTapped,
  physics: NeverScrollableScrollPhysics(),
  children: <Widget>[
    // 三个页面已经进行封装
    ListPage(
      noteDbHelpter: noteDbHelpter,
    ),
    CalendarPage(noteDbHelpter: noteDbHelpter),
    CenterPage(noteDbHelpter: noteDbHelpter),
  ],
)),
// 底部导航栏用 CupertinoTabBar 实现
bottomNavigationBar: CupertinoTabBar(
  // 导航集合
  items: <BottomNavigationBarItem>[
    BottomNavigationBarItem(
        activeIcon: Icon(
          Icons.event_note,
          color: Colors.blue[300],
        ),
        icon: Icon(Icons.event_note),
        title: Text('主页')),
    BottomNavigationBarItem(
        activeIcon: Icon(
          Icons.calendar_today,
```

```
                        color: Colors.blue[300],
                      ),
                      icon: Icon(Icons.calendar_today),
                      title: Text('日历')),
                  BottomNavigationBarItem(
                      activeIcon: Icon(
                        Icons.person,
                        color: Colors.blue[300],
                      ),
                      icon: Icon(Icons.person),
                      title: Text('个人中心')),
                ],
                currentIndex: _selectedIndex,
                onTap: setPageViewItemSelect,
              ),
            )),
          ));
    }

    void _onItemTapped(int index) {
      setState(() {
        _selectedIndex = index;
      });
    }

    // 底部单击切换
    void setPageViewItemSelect(int indexSelect) {
      _pageController.animateToPage(indexSelect,
          duration: const Duration(milliseconds: 300), curve: Curves.ease);
    }
}
```

以上是应用首页的整体框架及逻辑代码，其中涉及主页、日历、个人中心等页面，这些页面也需要编写代码实现，这里不再重复罗列，大家可以在 GitHub 上查看。

Flutter 备忘录应用中涉及剪贴板复制粘贴功能，实现比较简单，主要用到 Clipboard 和 ClipboardData 这两个组件，代码如下。

```
Clipboard.setData(
          ClipboardData(text: _noteList.elementAt(index).content));
```

备忘录中也用到了一些常用的时间工具类，用于将时间戳转换为指定格式，实现代码如下。

```
// DateTime 操作
class TimeUtils {
  static String getWeekday(int day) {
    switch (day) {
      case 1:
        return "一";
      case 2:
        return "二";
      case 3:
        return "三";
      case 4:
        return "四";
      case 5:
        return "五";
      case 6:
        return "六";
      case 7:
        return "日";
    }
  }

  static String getDateTime(DateTime dateTime) {
    return '${dateTime.year}年${dateTime.month}月${dateTime.day}日 ${dateTime.hour}:
${dateTime.minute}:${dateTime.second}';
  }

  static String getDate(DateTime dateTime) {
    return '${dateTime.year}年${dateTime.month}月';
  }
}
```

其他页面实现方式大同小异，其中备忘录通知刷新通过 EventBus 组件实现，之前的章节里讲解过用法，这里不再重复。增删改查功能通过数据库 sqflite 实现，这里也不重复讲解，直接给出数据库操作类代码，大家可以参考学习。

```
import 'package:flutter_note/entity/note.dart';
import 'package:sqflite/sqflite.dart';

// 数据库操作工具类
```

```
class NoteDbHelper {
  Database db;

  Future open(String path) async {
    // 打开/创建数据库
    db = await openDatabase(path, version: 1,
        onCreate: (Database db, int version) async {
      await db.execute(
          "create table notes (_id INTEGER primary key autoincrement,title TEXT not n
ull,content TEXT not null,star INTEGER not null,time INTEGER not null,weather INTEGE
R not null)");
      print("Table is created");
    });
  }

  Future<Database> getDatabase() async {
    Database database = await db;
    return database;
  }

  // 增加一条数据
  Future<Note> insert(Note note) async {
    note.id = await db.insert("notes", note.toMap());
    return note;
  }

  // 通过 ID 查询一条数据
  Future<Note> getNoteById(int id) async {
    List<Map> maps = await db.query('notes',
        columns: [
          columnId,
          columnTitle,
          columnContent,
          columnTime,
          columnStar,
          columnWeather
        ],
        where: '_id = ?',
        whereArgs: [id]);
    if (maps.length > 0) {
```

```
    return Note.fromMap(maps.first);
  }
  return null;
}

// 通过关键字查询数据
Future<List<Note>> getNoteByContent(String text) async {
  List<Note> _noteList = List();
  List<Map> maps = await db.query('notes',
      columns: [
        columnId,
        columnTitle,
        columnContent,
        columnTime,
        columnStar,
        columnWeather
      ],
      where: 'content like ? ORDER BY time ASC',
      whereArgs: ["%" + text + "%"]);
  if (maps.length > 0) {
    for (int i = 0; i < maps.length; i++) {
      _noteList.add(Note.fromMap(maps.elementAt(i)));
    }
    return _noteList;
  }
  return null;
}

// 通过 ID 删除一条数据
Future<int> deleteById(int id) async {
  return await db.delete('notes', where: '_id = ?', whereArgs: [id]);
}

// 更新数据
Future<int> update(Note note) async {
  return await db
      .update('notes', note.toMap(), where: '_id = ?', whereArgs: [note.id]);
}
```

```
  // 关闭数据库
  Future close() async => db.close();
}
```

以上代码涵盖了数据库增删改查基础方法，是备忘录应用中增删改查操作的基础。

以上我们完成了一个简单的备忘录开发，基本的功能都具备。在开发时需要先分析界面和需求功能，然后设计和选择架构、组件，通过这种方式开发的应用会具有更强的健壮性。完整的项目代码可以在 GitHub 上下载。

18.2　编写一个 TV 应用

我们知道，目前的智能电视和机顶盒基本是基于 Android 系统开发的，所以一般的 TV 应用也都是采用 Android 原生 API 开发的。Google 对于 Android TV 开发也制定了一些规范，同时推出了一些支持库。本节我们将拓展内容，尝试使用 Flutter 开发 TV 应用，介绍 TV 应用开发的主要难点：按键监听、焦点处理、焦点框效果处理。

由于 Google 官方并没有推出 TV 版的 Flutter SDK，所以用 Flutter 开发 TV 应用会遇到很多困难，主要涉及按键监听、焦点处理、焦点框效果处理。

原生的 Android 控件中默认包含焦点处理属性，可直接配置使用。而通过 Flutter 开发 TV 应用需要自己实现按键监听、焦点处理、焦点框效果处理，比较麻烦。如果能处理好这些问题，开发 TV 应用便没有什么难度。新版 Flutter 中多了 DefaultFocusTraversal 组件，我们可以借助这个组件指定方向移动焦点，简化开发。

18.2.1　按键监听

Flutter 组件能够监听遥控器或手机端按键事件的前提是，这个组件已经获取了焦点。实现按键监听需要使用 RawKeyboardListener 组件，其构造方法如下。

```
const RawKeyboardListener({
    Key key,
    @required this.focusNode,// 焦点节点
    @required this.onKey,// 核心，按键监听
    @required this.child,// 接收焦点的子组件
  })
```

在本例中，我们也是使用 RawKeyboardListener 组件来实现按键监听的，具体来说可以实现

遥控器的上、下、左、右等按键和焦点的监听，代码如下。

```
FocusNode focusNode0 = FocusNode();
... ...
RawKeyboardListener(
      focusNode: focusNode0,
      child: Container(
        decoration: getCircleDecoration(color0),
        child: Padding(
          child: Card(
            elevation: 5,
            shape: CircleBorder(),
            child: CircleAvatar(
              child: Text(''),
              backgroundImage: AssetImage("assets/icon_tv.png"),
              radius: radius,
            ),
          ),
          padding: EdgeInsets.all(padding),
        ),
      ),
      onKey: (RawKeyEvent event) {
        if (event is RawKeyDownEvent && event.data is RawKeyEventDataAndroid) {
          RawKeyDownEvent rawKeyDownEvent = event;
          RawKeyEventDataAndroid rawKeyEventDataAndroid = rawKeyDownEvent.data;
          print("keyCode: ${rawKeyEventDataAndroid.keyCode}");
          switch (rawKeyEventDataAndroid.keyCode) {
            case 19: // KEY_UP（上按键）
              FocusScope.of(context).requestFocus(_focusNode);
              break;
            case 20: // KEY_DOWN（下按键）
              break;
            case 21: // KEY_LEFT（左按键）
              FocusScope.of(context).requestFocus(focusNode4);
              break;
            case 22: // KEY_RIGHT（右按键）
              FocusScope.of(context).requestFocus(focusNode1);
              break;
            case 23: // KEY_CENTER（中间按键）
              break;
```

```
            case 66: // KEY_ENTER（确认键）
            break;
            default:
              break;
          }
        }
      },
    )
```

18.2.2　焦点处理

前面讲到，Flutter 组件能够实现按键监听的前提是已经获取焦点。获取焦点很重要，具体要借助 FocusScope 组件来实现。用 FocusScope 手动和自动获取焦点的代码如下。

```
// 手动获取焦点（指定组件手动获取焦点）
FocusScope.of(context).requestFocus(focusNode0);
// 自动获取焦点（Flutter 按照内部焦点顺序自动获取焦点）
FocusScope.of(context).autofocus(focusNode0);
```

获取焦点后就可以进行焦点处理了，处理焦点需要用到 FocusNode 组件。比如我们想进行焦点移动处理，需要通过 DefaultFocusTraversal 组件指定方向，搜索下一个焦点，代码如下。

```
FocusScope.of(context)
                    .focusInDirection(TraversalDirection.up);
// 或者
DefaultFocusTraversal.of(context).inDirection(
                    FocusScope.of(context).focusedChild, TraversalDirection.up);

DefaultFocusTraversal.of(context)
                    .inDirection(_focusNode, TraversalDirection.right);
```

DefaultFocusTraversal 组件支持在上、下、左、右四个方向移动焦点，如果想手动指定下一个焦点，只需传入要处理焦点的 FocusNode 组件即可，代码如下。

```
FocusScope.of(context).requestFocus(focusNode);
```

18.2.3　焦点框效果处理

实现焦点框效果（如选中显示效果和隐藏效果）的主要原理是，利用 Flutter 的边框组件动态设置边框颜色、边框宽度、边框装饰。举例来讲，选中后可以将焦点框颜色设置为黄色，未选中时就设置为透明。

我们可以在最外层的 Container 里设置 BoxDecoration 属性，实现焦点框效果。

```
var default_decoration = BoxDecoration(
    border: Border.all(width: 3, color: Colors.deepOrange),
    borderRadius: BorderRadius.all(
      Radius.circular(5),
    ));
... ...
child: Container(
          margin: EdgeInsets.all(8),
          decoration: default_decoration,
          child: widget.child,
        ));
```

设置好边框效果后，接下来就可以编写项目了。这里只给出部分核心代码，完整代码可以在 GitHub 上下载查看。首先要绘制欢迎页面，这一步并不复杂，需要注意的是，TV 应用需要设置为横屏，并且页面上要有一个倒计时跳转到首页的功能，具体代码如下。

```
// 启动欢迎页面
import 'dart:async';

import 'package:flutter/material.dart';
import 'package:flutter/services.dart';

import 'ui/tv_page.dart';

void main() => runApp(MyApp());

class MyApp extends StatelessWidget {
  @override
  Widget build(BuildContext context) {
    SystemChrome.setEnabledSystemUIOverlays([]);
    // 强制横屏
    SystemChrome.setPreferredOrientations([
      DeviceOrientation.landscapeLeft,
      DeviceOrientation.landscapeRight
    ]);
    return MaterialApp(
      title: 'Flutter TV',
      debugShowCheckedModeBanner: false,
```

```
      theme: ThemeData(
        primarySwatch: Colors.blue,
      ),
      home: MyHomePage(),
    );
  }
}

class MyHomePage extends StatefulWidget {
  @override
  _MyHomePageState createState() => _MyHomePageState();
}

class _MyHomePageState extends State<MyHomePage> {
  Timer timer;

  @override
  void initState() {
    startTimeout();
    super.initState();
  }

  @override
  Widget build(BuildContext context) {
    return Scaffold(
      primary: true,
      backgroundColor: Colors.black54,
      body: Center(
        child: Text(
          '芒果 TV',
          style: TextStyle(
              fontSize: 50,
              color: Colors.deepOrange,
              fontWeight: FontWeight.normal),
        ),
      ),
    );
  }
```

```
  _toPage() {
    Navigator.pushAndRemoveUntil(
      context,
      MaterialPageRoute(builder: (context) => TVPage()),
      (route) => route == null,
    );
  }

  // 倒计时处理
  static const timeout = const Duration(seconds: 3);

  startTimeout() {
    timer = Timer(timeout, handleTimeout);
    return timer;
  }

  void handleTimeout() {
    _toPage();
  }

  @override
  void dispose() {
    if (timer != null) {
      timer.cancel();
      timer = null;
    }
    super.dispose();
  }
}
```

运行以上代码，得到欢迎页面，效果如图 18-4 所示。

图 18-4　欢迎页面效果

接下来要实现应用首页，应用首页界面也不复杂，只需使用 Scaffold 脚手架配合其他组件进行绘制即可完成。由于逻辑简单，这里不给出相关代码，各位读者可以到 GitHub 上下载查看。应用首页效果如图 18-5 所示。

图 18-5　应用首页效果

相比于实现应用首页，焦点框的处理比较复杂，这里给出封装好的核心焦点框处理代码，如下。

```
import 'package:flutter/material.dart';
import 'package:flutter/services.dart';
import 'package:flutter/widgets.dart';
```

```
class TVWidget extends StatefulWidget {
  TVWidget(
      {Key key,
      @required this.child,
      @required this.focusChange,
      @required this.onclick,
      @required this.decoration,
      @required this.hasDecoration = true,
      @required this.requestFocus = false})
      : super(key: key);

  Widget child;
  onFocusChange focusChange;
  onClick onclick;
  bool requestFocus;
  BoxDecoration decoration;
  bool hasDecoration;

  @override
  State<StatefulWidget> createState() {
    return TVWidgetState();
  }
}

typedef void onFocusChange(bool hasFocus);
typedef void onClick();

class TVWidgetState extends State<TVWidget> {
  FocusNode _focusNode;
  bool init = false;
  var default_decoration = BoxDecoration(
      border: Border.all(width: 3, color: Colors.deepOrange),
      borderRadius: BorderRadius.all(
        Radius.circular(5),
      ));
  var decoration = null;

  @override
  void initState() {
```

```
    super.initState();
    _focusNode = FocusNode();
    _focusNode.addListener(() {
      if (widget.focusChange != null) {
        widget.focusChange(_focusNode.hasFocus);
      }
      if (_focusNode.hasFocus) {
        setState(() {
          if (widget.hasDecoration) {
            decoration = widget.decoration == null
                ? default_decoration
                : widget.decoration;
          }
        });
      } else {
        setState(() {
          decoration = null;
        });
      }
    });
  }

  @override
  Widget build(BuildContext context) {
    if (widget.requestFocus && !init) {
      FocusScope.of(context).requestFocus(_focusNode);
      init = true;
}
    // 监听按键
    return RawKeyboardListener(
        focusNode: _focusNode,
        onKey: (event) {
          if (event is RawKeyDownEvent &&
              event.data is RawKeyEventDataAndroid) {
            RawKeyDownEvent rawKeyDownEvent = event;
            RawKeyEventDataAndroid rawKeyEventDataAndroid =
                rawKeyDownEvent.data;
            print("keyCode: ${rawKeyEventDataAndroid.keyCode}");
            switch (rawKeyEventDataAndroid.keyCode) {
```

```
          case 19: // KEY_UP（上按键）
          // DefaultFocusTraversal.of(context).inDirection(
          // FocusScope.of(context).focusedChild, TraversalDirection.up);
            FocusScope.of(context)
                .focusInDirection(TraversalDirection.up);
            break;
          case 20: // KEY_DOWN（下按键）
            FocusScope.of(context)
                .focusInDirection(TraversalDirection.down);
            break;
          case 21: // KEY_LEFT（左按键）
            // FocusScope.of(context).requestFocus(focusNodeB0);
            FocusScope.of(context)
                .focusInDirection(TraversalDirection.left);
            // 手动指定下一个焦点
           // FocusScope.of(context).requestFocus(focusNode);
            break;
          case 22: // KEY_RIGHT（右按键）
            FocusScope.of(context)
                .focusInDirection(TraversalDirection.right);
            break;
          case 23: // KEY_CENTER（中心按键）
            widget.onclick();
            break;
          case 66: // KEY_ENTER（确认键）
            widget.onclick();
            break;
          default:
            break;
        }
      }
    },
    child: Container(
      margin: EdgeInsets.all(8),
      decoration: decoration,
      child: widget.child,
    ));
  }
}
```

运行以上代码，效果如图 18-6 所示。

图 18-6 焦点框效果

通过图 18-6 可以看到，选中的焦点（左上角）外围有了焦点框，这样便于我们在操作遥控器时明确当前选中的位置在哪里。焦点框效果处理是 TV 应用开发中常用的知识点，各位读者应掌握。

关于 Flutter TV 应用开发的内容就讲解这么多，要想查看完整示例代码，可以在 GitHub 上搜索 flutter_tv 并下载、查看。